Fundamental Engineering Mechanics

Dr Peter J Ogrodnik
Senior Lecturer, School of Engineering,
Staffordshire University

Fundamental Engineering
Mechanics

An imprint of **Pearson Education**

Harlow, England · London · New York · Reading, Massachusetts · San Francisco
Toronto · Don Mills, Ontario · Sydney · Tokyo · Singapore · Hong Kong · Seoul
Taipei · Cape Town · Madrid · Mexico City · Amsterdam · Munich · Paris · Milan

Pearson Education Limited
Edinburgh Gate
Harlow
Essex CM20 2JE
England

and Associated Companies throught the world

Visit us on the World Wide Web at:
http://www.pearsoneduc.com

First published 1997

British Library Cataloguing in Publication Data
A catalogue entry for this title is available from the British Library

ISBN 0-582-29799-0

Library of Congress Cataloging-in-Publication Data
A catalog entry for this title is available from the Library of Congress
Transferred to digital print on demand, 2006
Printed and bound by Antony Rowe Ltd, Eastbourne

This book is dedicated
to my Mum and Dad.

Throughout their lives they have had to fight,
for their freedom and for their health;
they deserve this prize.

Contents

Chapter 10 **Case studies**

Appendices

Preface

What is engineering mechanics? Before we can answer this question we need to understand a little more about the engineering profession and the work of an engineer.

The word *engineer* can be both a noun or a verb, both having roughly the same meaning. It comes from the Latin *ingenium* or clever invention. It may also be attributed to *ingenious*. The noun is therefore a name attributed to a profession which by its nature means 'to be ingenious' or 'to produce clever inventions'. As a verb, *engineer* has been used to describe military, political and social tactics as in 'to engineer a situation'. In other words, 'to use all resources available with intelligence, courage and wit to achieve an outcome which is desired.' This is how I see engineering, and most practising engineers would agree that their basic function is to provide the best solution to a theoretical or practical problem using whatever is available to them. However, we have the further restriction that the environment cannot withstand wholesale 'plundering', so whatever solutions we arrive at must make best use of available resources, and not waste them.

Engineers have to live with their mistakes. It is one of the few professions where a mistake can directly lead to a loss of life, and this is a very great responsibility to bear. Upon completion of the Menai Bridge, Stephenson is reputed to have said, 'At last I can sleep'. Unfortunately, a major portion of an engineer's training is directed to understanding all the things that can go wrong with a system. One example that is often mentioned is the answer to the problem $1 + 1 = x$, clearly mathematics states that x must be equal to 2. This statement is fine when we deal with whole objects. If we were adding the lengths of two boards, each 1 m long, an engineer would say that x is about 2 m. This is because an engineer knows that, in real life, nothing is exact! This understanding is a double-edged sword; on one hand we hope that all eventualities have been considered, on the other we know that some are beyond our imagination. An engineer must be able to visualise, imagine and model the performance of an artefact during the design stage; it is too late to go back once the object has been made and sold!

A firm underpinning of theoretical analysis is a cornerstone of engineering, another is experience. It is important that the engineer knows which equation solves which particular problem, but it is far more important to be able to identify the problem in the first instance. This ability to look at a real-life situation and to simplify it to basic engineering models is what distinguishes engineers from other professionals. However, an engineer also needs to be able to model, and the language of modelling is mathematics. An engineer is a weird combination of mathematician, physical scientist and artisan who retains a healthy scepticism.

It is hard to define the first engineer, but feats such as the building of the pyramids in Egypt, or the Great Wall of China or even Roman roads are all classed as great engineering achievements. A well-documented example features Hero of Alexandria, who invented a simple precursor to the modern steam turbine, nowadays used in most power stations. Imagine what it must have been like for an Alexandrine to come face to face with a spinning object spouting steam!

Throughout history major changes in the standard of living can be attributed to engineering, none more so than in the past few hundred years. We only need to look around the home to realise the impact that engineering has had on our modern lives, and we only have to look outside to see what effect we have had on our environment! Thus engineers have some onerous responsibilities. We have to satisfy customer demand and we must guard against wanton destruction of the environment.

So, back to our original question, What is engineering mechanics? The use of the term *engineering* clearly distinguishes it from other professions, and mechanics is often thought to be the oldest science. We have established that, as engineers, we need to be able to model real-life situations; sometimes we need to be able to take a real-life situation and produce a model so abstract that it would compare with a Picasso. One important branch of the modelling process is to be able to understand the mechanics of a system, or how the system works. In terms of a single object, this may be how it is moving. Clearly the more complex the object, the more we need to model. As engineers, we must understand the language of the modelling process and build up a toolbox of modelling tools; just as an artist would build up a palette of colours. Mathematics is one tool, but this tool is useless without the ability to be able to model in the first instance. Always in the back of our minds we must remember our objective: to produce a product (whatever the product may be) which is fit for its purpose but which is not wasteful of resources and which has limited impact on the surrounding environment.

It is with this in mind that the text has been prepared. Students who are beginning their studies in engineering and technology need a firm foundation on which to base their modelling skills. It aims to bring together within one text the fundamental analytical and conceptual models used in engineering mechanics.

The first chapters introduce fundamental analytical models from basic units through to energy. Subsequent chapters build on this knowledge and apply these fundamental models to important branches of engineering mechanics. The case studies in the final chapter consolidate the knowledge base and develop the student's ability to take a real-life situation and produce an abstract model. The case studies demonstrate how the understanding of some simple basic rules enables engineers to be able to model quite complex systems. Each chapter has tutorial problems which the student should complete *before* attempting the next chapter. This is important if the student is to benefit from the structure of the text. The significant equations, definitions and statements have been highlighted throughout, and should be memorised so they can be recalled at will.

As an aid to revision, Appendix E contains several past examination questions from the engineering foundation course of Staffordshire University. You should attempt these questions under conditions as similar as possible to a real examination. This will help you to become accustomed to the rigours of the examination procedure itself. At the end of the day, only you can learn a subject, no one else can learn it for you. So give it a go, you have nothing to lose, but everything to gain.

If this text has enabled a few more people to start a career in engineering and technology then I am very happy. Best of luck, and enjoy your studies.

To complete the text there is also a World-Wide Web page at http://www.awl-he.com which contains accompanying software and self-assessment routines. The home page has full details of all the facilities available. If you have any comments on the book or find any errors, please email them to this website. Academic staff may be interested in the teaching material posted there

Finally, the preparation of a textbook goes through many phases, one of which is being 'torn apart' by friends and colleagues. To this end, I wish to thank the following people for their support advice and comments; Tarik Al-Shemmeri, Barrie Delves, Graham Hourouge, Ian Moorcroft, Peter Thomas and Roy Wood. I also wish to thank all the numerous foundation students who have suffered my awful jokes but who also supplied the initial impetus. Thanks also go to Toni Salwey who helped with the typing and who constructed the index.

My family – Mum, Dad and the rest of the clan – have supported me throughout my career. Without their support, this book would not exist. Thanks with all my heart. Finally two very special people deserve the biggest thanks, because they hardly saw me whilst I was writing this book, Natasha and Lynda Ogrodnik.

Peter J. Ogrodnik

Acknowledgements

The author wishes to thank the following organisations for their contribution to the text: Denison Mayes Group for Figure 9.5 and for the data used in Section 10.4, and Staffordshire University for their permission to use past examination questions and for their support during the development of the text.

Notation

All symbols in bold characters denote vector quantities; those in italics denote scalar quantities.

$\mathbf{a}$	linear acceleration, m/s^2
$\mathbf{a}_c$	linear acceleration, m/s^2
A	cross-sectional area, m^2
c	specific heat capacity, J/kg.K
e	coefficient of restitution
E	modulus of elasticity (Young's modulus), N/m^2
E_c	stored elastic energy, J
E_k	kinetic energy, J
E_p	potential energy, J
$\mathbf{F}$	force, N
$\mathbf{F}_b$	buoyancy force, N
$\mathbf{F}_c$	centrifugal force, N
$\mathbf{F}_T$	tension in a rope, cable or spare, N
$\mathbf{g}$	gravitational acceleration, m/s^2
G	universal gravitation constant
G	modulus of rigidity, N/m^2
h	height, m
$\mathbf{i}$	unit vector
I	second moment of area, m^4
I	moment of inertia, kgm^2
$\mathbf{j}$	unit vector
k	stiffness, N/m
l	beam length, m
L	length, m
L	angular momentum, Nms
m	mass, kg
$\dot{m}$	mass flow rate, kg/s
$\mathbf{M}$	moment, Nm
p	pressure, N/m^2
p_0	atmospheric pressure, N/m^2

$\mathbf{p}$	momentum, Ns
P	power, W
Q	heat, J
r	radius, m
$\mathbf{R}$	reaction force, N
R	gas constant, J/kg.K
s	distance, m
SF	safety factor
t	time, s
t	relative temperature, °C
T	absolute temperature, K
T	torque, Nm
$\mathbf{u}, \mathbf{v}$	velocity, m/s
$\mathbf{v}_1$	initial velocity
$\mathbf{v}_2$	final velocity
V	volume m^3
$\dot{V}$	volumetric flow rate, m^3/s
$\mathbf{x}, \mathbf{y}, \mathbf{z}$	displacement in x, y or z axes, m
Z	height above a datum, m
α	angular acceleration, rad/s^2
α	coefficient of thermal expansion strain/K
γ	shear strain
ε	direct strain
θ	angular displacement, rad
π	ratio of a circle's circumference to its diameter.
ρ	density, kg/m^3
σ	direct stress, N/m^2
σ_y	yield stress, N/m^2
$\sigma_{0.1\%}$	0.1% proof stress, N/m^2
τ	shear stress, N/m^2
ω	angular velocity, rad/s

Mathematical symbols

$=$	equal to
$\neq$	not equal to
$\approx$	approximately equal to
$\propto$	proportional to
$+$	addition
$-$	subtraction
$\times$	multiplication
$/$	division
$<$	less than
$>$	greater than
∞	infinity
$\sum$	sum of

Units

In this chapter we shall be examining the fundamental physical quantities associated with the discipline of engineering mechanics. The differences between vectors and scalars will be explored and the way they are handled in calculations will be introduced. At this stage it is important to stress that a great deal of engineering mechanics is based on vector algebra so the student should become familiar with this section. Failure to do so before going to subsequent chapters may cause confusion.

Unfortunately, although metric units have been around for many years, a large proportion of the engineering industry still works with the old imperial units. Thus the final section introduces some conversion factors so that when you finally reach industry you will not fall foul of the mismatch between SI and imperial units (as they are known).

1.1 SI Units

All written languages have a character set which has to be learnt in order to understand the written text, this is also the case with engineering mechanics. Since the calculations of an engineer are international, so the language is international. The use of mathematics is universal, but to make the calculations and theorems intelligible the symbols used must also be universal. This is particulalry important when considering the international nature of engineering. To overcome this problem the SI (Système International) unit was developed (and is still developing). All physical quantities have been given a unique name and a unique symbol under this system; in reality the symbols are nothing more than an agreed shorthand. Later in your studies you will meet equations which have many separate symbols, without this agreed shorthand the equations would become unintelligible! This has the added benefit that a French engineer can understand the calculations of a British engineer, even though they may not be able to speak to one another.

The system also standardised the units themselves such that we have reproducible datums. Take, for instance, the unit of length. Through history it has been based on human dimensions. The Egyptians used cubits (elbow to fingertips). In England barleycorns were used by cobblers to measure the size of the foot. Edward I extended this idea in 1305 and stated that three barleycorns end-to-end were an inch. Henry I went further and stated that the standard yard was the distance from his nose to his fingertips, which was duly measured and an iron bar forged. Following monarchs were of different build and had different whims, hence this 'standard yard' was not stable! In the sixteenth century the foot was standardised in Germany using the average of 16 men's left feet. The French revolution paved the way for science, instead of monarchs, to define length. The French scientists decided that the 'metre' should be based on the earth, 1/10 000 000 of the distance from the North Pole to the equator, to be exact. This they estimated by surveying the distance from Dunkirk to Barcelona. Not to be outdone, the British (after the Palace of Westminster was burned down in 1834 and with it our 'standard yard') also tried science to define the yard. First the yard was defined as the length of a pendulum whose swing took one second, but this was impossible to reproduce experimentally. Instead they took an average of all the copies of the old standard yard. Special 'standard yard' sticks were manufactured and sent over the world, only to be found to be changing all the time! Clearly not a standard of measurements!

SI uses light as the datum for length, one metre being the distance travelled by light in 1/299 792 458 of a second. A platinum–iridium (which is very stable) cylinder in Sevres, France, is used as the universal datum.[1] This datum enables engineers and scientists to communicate dimensions across the world and, more important, to reproduce them.

The use of symbols covered by SI are dictated under ISO 1000/BS 5555 and ISO 30/BS 5775. Table 1.1 illustrates some physical quantities, in particular those we shall meet regularly in this textbook. This table is a starting point, it is worthwhile for you to construct your own version. Whenever you meet a new quantity you can add it to your table and start to familiarise yourself with its correct symbol. Each quantity has a symbol and a unit. As stated before, the symbol is the shorthand and the unit is the

Table 1.1 Fundamental SI units

Quantity	Symbol	Unit	Definition
Mass	m	kg (kilogramme)	1 kg is equal to the mass of a reference international prototype held in Sevres, France
Length	l	m (metre) or mm (millimetre)	1 metre is the distance travelled by light in a vacuum in 1/299 792 458 of a second
Time	t	s (second)	1 second is the duration of 9 192 631 770 periods of the radiation corresponding to the transition between the two hyperfine levels of the ground state of caesium 133
Temperature	T	K (kelvin)	1 kelvin is 1/273.16 of the triple point of water at sea level

[1] Windfall Films have made the series *Measure for Measure*; this is recommended viewing for those interested in the history of measurement.

value of the quantity with respect to the datum given by the defintion. Four quantities are given; they are the fundamental quantities from which we can define all others. Note that the unit of length is the metre or millimetre, do not use centimetre even though every ruler in the country has them marked!

In Table 1.1 you will have noticed two items of shorthand which may be unfamiliar (even though you may use them every day). These are *kilo* (as in kilogramme) and *milli* (as in millimetre). This shorthand has been adopted to cope with the massive range of scales engineers and scientists have to deal with. It is based on scales of 1000 – each new unit is 1000 times greater than the one before. Table 1.2 details the preferred units, their name, their shorthand and their value in scientific notation and exponential notation (used by computers and calculators).

Table 1.2 Preferred multiples of units

Value	Shorthand (prefix)	Name	Scientific notation	Exponential notation
1/1 000 000 000 000	p	pico	$\times 10^{-12}$	E − 12
1/1 000 000 000	n	nano	$\times 10^{-9}$	E − 09
1/1 000 000	μ	micro	$\times 10^{-6}$	E − 06
1/1000	m	milli	$\times 10^{-3}$	E − 03
1			1	E00
1000	k	kilo	$\times 10^{3}$	E + 03
1 000 000	M	mega	$\times 10^{6}$	E + 06
1 000 000 000	G	giga	$\times 10^{9}$	E + 09
1 000 000 000 000	T	tera	$\times 10^{12}$	E + 12

As you can see, from the first column of Table 1.2, one soon runs out of room when dealing with engineering numbers. The way in which Table 1.2 is used can be illustrated by the following example.

The nominal diameter of a human hair is approximately 0.000 075 m. This number can be written in three ways, all of which mean the same

0.075 mm   (which means 0.075/1000 m)
75 μm   (which means 75/1 000 000 m)
75×10^{-6} m (which means 75/1 000 000 m)

but note how the information conveyed is easier to interpret. We do not need to count the zeros!

Note how the system is used, the scientific notation is used as a multiple of the number, as in 75×10^{-6} m. The shorthand is used as a prefix to the unit's symbol itself, as in 0.075 mm or 75 μm.

It takes practice to use this system, but modern hand-held calculators have an ENG button which converts numbers to these units for us. For an engineer it is important to use these 'orders of scale' because they give us a 'feel' for the number, for its scale. Often the actual value of a number is a side issue. It is the scale (i.e. k, M or G) which sends alarm bells ringing in our minds.

1.2   Coordinate systems

In this textbook we will be concerned with the position and motion of bodies in space. Hence we need to be able to describe both position and motion. To do this sucessfully we have to work to a recognised coordinate system. In practice most engineers commonly use *rectangular* and *polar* coordinate systems. Luckily, once you have them mastered, it is possible to swap between both systems with ease. The two systems are required because the mathematics of a system can be simplified by using one or the other. Nowadays, pocket calculators do all the calculations for you, so there really is no excuse for avoiding them!

Figure 1.1 illustrates an object in space; it is immaterial what the object is since this method is applicable to any object. Firstly, it must be recognised that we must set a datum, a point from which all measurements are taken. Since we are in a three-dimensional world we tend to measure an object's position in 3D. On the figure, three axes have been drawn to represent the three dimensions; these are normally called x, y and z. Note the orientation between these axes; if this is not adhered to, things can go drastically wrong. The axes are often called orthogonal axes (because they are at right angles, 90″, to one another). The arrowheads indicate the + (positive) direction (forwards if you wish), the opposite direction is called the − (negative) direction.

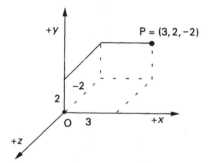

Figure 1.1   Rectangular coordinate system

The intersection between the three axes x, y and z is used as the datum, often called the origin. One thing which is often missed is that the position of this intersection is quite literally up to us. We can put it *anywhere we choose*. Put the origin in the right place and things can be made very easy; put it in the wrong place and things can become very difficult. Engineering has a lot to do with positioning datums and origins in the right place, using experience and common sense. Note that rectangular coordinate systems use the *roman alphabet*.

Let us now consider an object which is positioned at a point in space P. How can we describe its position relative to the origin O? An examination of the figure shows that P is +3 along the x-axis from O. Along the y-axis it is +2 units and along z, −2 units. This is clearly a very long way of describing position, to make it more concise we use a special form of notation:

P = (position in *x*, position in *y*, position in *z*)relative to the origin O

or

P = (3, 2, −2)

The polar coordinate system[2] is based on one orthogonal plane and an origin. The polar system is based on the concept of a line of length *r* being able to rotate around the origin O of a fixed axis *x*, *y* or *z* as illustrated in Figure 1.2(a). In this figure the line rotates about the *z*-axis and as such rotates within the *x*–*y* plane.

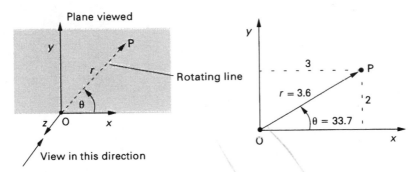

Figure 1.2 Polar coordinate system: (a) three-dimensional Cartesian axes and (b) two-dimensional polar axes

Figure 1.2(b) is the same diagram as Figure 1.2(a) but drawn in 2D (viewed in the direction shown, so the *z*-axis sticks out of the page). In fact what we view is a 'plane' corresponding to a pair of orthogonal axes *x* and *y*. Viewing along the *x*-axis we see the *y*–*z* plane, and viewing along the *y*-axis we see the *x*–*z* plane.

To define P in polar coordinates we draw an imaginary line between O and P; P is then defined by the length of this line and its angular orientation with respect to the *x*-axis. Thus the position of P is really defined by the line OP, where the length of OP is *r* = 3.6 and its orientation with respect to the *x*-axis is θ = 33.7°. For now this can be estimated using a ruler and protractor, but later we shall be using more accurate methods.

Note that a positive angle is anticlockwise with respect to the *x*-axis. Furthermore, angular orientation is always denoted by a Greek character. Strictly speaking, the angle θ should be measured in radians, but we shall meet this in greater detail later.

Figure 1.2 suggests there is a link between rectangular coordinates and polar cordinates. The link is relatively simple; we shall be meeting the method of conversion when we examine vector algebra in Section 1.4.

---

[2] Polar coordinates are a 2D system. They can be 3D, then they are called *spherical coordinates*.

1.3 Scalars and vectors

Within the field of engineering mechanics we meet two types of physical quantity, one is a *scalar* quantity the other is a *vector* quantity.

A scalar quantity only requires a magnitude to describe it fully. Examples of scalar quantities are distance and temperature.

A vector quantity requires both magnitude and direction to describe it fully. Examples of a vector quantity are force and velocity.

To exemplify the important difference between these two quantities, let us consider someone asking directions for a tourist location, a famous castle, from two people in the centre of a small town (Figure 1.3 illustrates a map of the tourist's location).

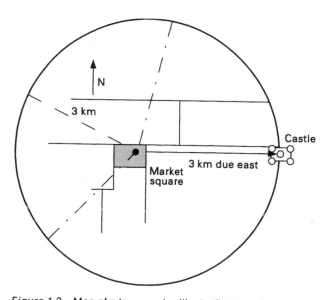

Figure 1.3   Map of a town centre illustrating a scalar and a vector

The first person asked states that the castle is about 3 km away from where they stand. This statement gives the tourist the distance but no direction, thus the castle can lie anywhere on the circle (as illustrated by the three dotted lines on Figure 1.3). The information given was scalar.

The tourist stops another person, who agrees that the castle is 3 km away but is also due east from where they stand. This information contains both magnitude and direction and is thus a vector. It also locates the castle for the tourist and enables him and his family to have an enjoyable afternoon.

The story above conveys the importance of knowing the difference between scalar and vector quantities. Later we shall see that we have to treat vectors differently to scalars; for instance, vector addition is different from the addition of scalars. For this reason, Section 1.4 is solely concerned with vector algebra.

Example 1.1   *From the list below determine which of the physical quantities are scalars and which are vectors:*

(a) *temperature*
(b) *length*
(c) *area*
(d) *weight*

SOLUTION
Items (a), (b) and (c) are scalars.
Item (d) is a vector. Why? Because weight always acts down, it has direction.

1.4 Vector algebra

Vectors are the basis of most engineering mechanics, and vector analysis uses different methods to scalar analysis. For the purposes of this text, we require four main calculations:

(i)   Addition and subtraction of vectors
(ii) Determination of the resultant vector from (i)
(iii) Conversion of a vector given in polar coordinates to rectangular coordinates
(iv) Conversion of a vector given in rectangular coordinates to polar coordinates

1.4.1 Describing terms as vectors
Before we can approach vector algebra, we have to describe a vector mathematically. Figure 1.4 illustrates two points located at A and B.

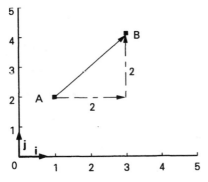

Figure 1.4   Determining an expression for the vector that describes moving from A to B

From Section 1.2 we can determine the positions of A and B in rectangular coordinates, remembering that to describe a point we use $P = $ (position in x, position in y):

A $= (1, 2)$
B $= (3, 4)$

Let us now consider a scenario where we are standing at point A and where we walk in a straight line to point B. How would we describe this movement? To describe it fully we need the magnitude, to describe how far we walk, and we need the direction, thus we require a vector. The bold arrow on Figure 1.4 illustrates this. Its length describes the distance walked (the magnitude) and its angular orientation with respect to the *x*-axis represents the direction.

Vectors are special quantities which have their own notation. Firstly we need to be able to distinguish between vectors and scalars in equations. There are two common ways to do this: by writing the symbol in bold or by drawing an arrow above the symbol, **a** or $\vec{a}$. We shall be using bold characters to indicate vector quantities.

We can call the vector that indicates moving from A to B whatever we like, but we must make its name sensible so that others can understand what we are doing. As you work through this text you will meet alternative ways of naming vectors. In this instance we shall call the vector **ab**, since we are describing motion from A to B. There is no significance of the change in case (upper to lower). To determine **ab**, we just need to consider the difference between A and B.

Table 1.3 shows that to determine the vector which describes moving from A to B one simply subtracts the position at A from the position at B. The motion actually

Table 1.3 Determining a vector from two points

	x	*y*
B	3	4
A	1	2
B − A	(3 − 1) = 2	(4 − 2) = 2

involves moving a total of +2 in the *x*-direction and +2 in the *y*-direction.

In practice we use shorthand for vectors. In rectangular coordinates we use *unit vectors* to develop a shorthand. Unit vectors are vectors which are 1 unit long and they point in the +*x* (**i**) and +*y* (**j**) directions. Thus 2**i** is shorthand for '2 units in the *x*-direction', and similarly 2**j** means '2 units in the *y*-direction'. Hence vector **ab** can be written as

$$\mathbf{ab} = 2\mathbf{i} + 2\mathbf{j}$$

Being able to separate a vector into its *x* and *y* components is a very powerful tool; it helps when adding vectors but is particularly powerful in engineering mechanics problems. And it is immaterial whether we move in the *x* or *y* direction first. The answer is still the same (as illustrated by the light arrows on Figure 1.4). It is also important to realise that the component vectors are not separated in space. The vectors **ab**, 2**i** and 2**j** all start from the same point. So when moving from A to B, the vectors 2**i** and 2**j** start and finish *at the same time*.

Example 1.2 *If* A $= (4, 5)$ *sketch point B if* **ab** $= -3\mathbf{i} - 2\mathbf{j}$. *If the motion was from B to A what would the vector describing this motion be?*

SOLUTION
First let us consider what the vector **ab** means.

ab $= -3\mathbf{i} - 2\mathbf{j}$ means 'move -3 in the x-direction and -2 in the y-direction'. In numerical form this can be expressed as

	x	y
A	4	5
ab	-3	-2
B	$4 - 3$	$5 - 2$
	$= 1$	$= 3$

Hence B $= (1, 3)$. In graphical form this can be represented by Figure E1.2.

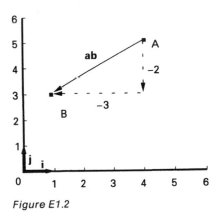

Figure E1.2

To determine the vector describing movement from B to A, we do the opposite of moving from A to B. Hence

ba $= -(\mathbf{ab}) = 3\mathbf{i} + 2\mathbf{j}$

1.4.2   Vector addition
The addition of vectors is simplified when the vector is described using rectangular coordinates. Consider the street plan illustrated in Figure 1.5. If a person lives in the house at A and wants to take a walk to the local shop at D what path can they take? One possible route is indicated by the dotted arrows; it involves three different vectors **ab**, **bc** and **cd**, which correspond to walking from A to B, B to C and C to D respectively.

There are two ways to add vectors: one is graphical, the other numerical. The graphical method involves drawing vectors to scale on a sheet of paper; addition is then

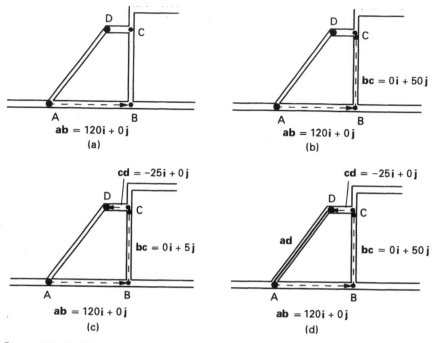

Figure 1.5   Addition of vectors

carried out by drawing vectors end-to-end as illustrated by Figure 1.5(a), (b) and (c). The end result of the addition of the three vectors is that the person walked from A to D. When two or more vectors are added the answer is called the *resultant vector*. Thus the addition of vectors **ab**, **bc** and **cd** is the resultant vector **ad**. If we draw the vectors to scale, on graph paper, the resultant vector can be determined by measurement.

Numerically, since the vector has been described in rectangular coordinates, the addition of the vectors is simply the addition of the x and y components as separate entities, therefore the resultant vector **ad** is described by

$$\mathbf{ad} = 95\mathbf{i} + 50\mathbf{j}$$

Which is the same as the vector describing the person walking directly from A to D. This process is illustrated in Table 1.4.

Table 1.4   Determining a resultant vector by vector addition

Vector	x	y
ab	120i	0j
bc	0i	50j
cd	−25i	0j
Resultant **ad**	95i	50j

To subtract vectors numerically one performs the same task as detailed in Table 1.4, but instead of adding the components, they are subtracted:

ab − **bc** = **ab** + (−**bc**)

Graphically the method of solution is slightly different to that presented earlier. We can draw the vectors starting from the same point; the difference between the two vectors is the vector drawn from the end of **bc** to the end of **ab** (or we can add (−**bc**) to **ab**). Both methods are illustrated in Figure 1.6. Clearly **ab** − **bc** = 120**i** − 50**j** (prove this for yourself using both methods).

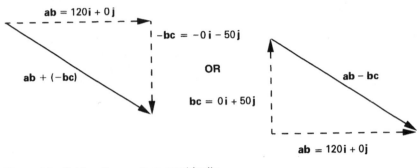

Figure 1.6   Subtracting vectors graphically

Example 1.3

(a) *Sketch the following points on a diagram depicting the* x–y *plane.*

$$A = (3, 4), B = (3, 6), C = (1, 1)$$

(b) *Determine the vectors which represent moving from*
  *A to B (**ab**)*
  *B to C (**bc**)*
  *A to C (**ac**)*
(c) *Show that* **ac** = **bc** + **ab**.

SOLUTION
(a) The x–y plane is the plane viewed such that z points out of the page. The resulting points lie such that A is +3 units along x and +4 units along y; B is +3 units along x and +6 units along y; C is +1 units along x and y (as shown in Figure E1.3)
(b) Moving from A to B (**ab**)

	x	y
B	3	6
A	3	4
ab	0**i**	+2**j**

hence **ab** = 2**j**.

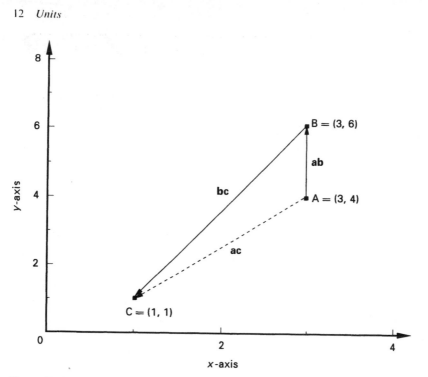

Figure E1.3

Moving from B to C

	x	y
C	1	1
B	3	6
bc	$-2i$	$-5j$

hence **bc** $= -2i - 5j$.

Moving from A to C

	x	y
C	1	1
A	3	4
ac	$-2i$	$-3j$

hence **ac** $= -2i - 3j$.

The vectors **ab**, **bc** and **ac** are shown on Figure E1.3

(c) Moving from A to C (**ac**)

$$\mathbf{ac} = -2i - 3j$$
$$\mathbf{ab} + \mathbf{bc} = (0 - 2)i + (2 - 5)j = -2i - 3j = \mathbf{ac}$$

Example 1.4   *Using the vectors* **ab**, **bc** *and* **cd** *from Example 1.3, show that* **ab** = **bc** − **ac**.

SOLUTION

This problem may be approached using two methods, vector diagrams and vector algebra.
   Using vector algebra we note (from Section 1.4.2)

$$\mathbf{ab} = \mathbf{ac} - \mathbf{bc} = \mathbf{ac} + (-\mathbf{bc})$$

hence

$$\begin{aligned}
\mathbf{ab} &= -2\mathbf{i} - 3\mathbf{j} + -(-2\mathbf{i} - 5\mathbf{j}) \\
&= (-2 + 2)\mathbf{i} + (-3 + 5)\mathbf{j} \\
&= 0\mathbf{i} + 2\mathbf{j}
\end{aligned}$$

Graphically this is solved by determining the vectorial difference between **ac** and **bc** as shown in Figure E1.4.

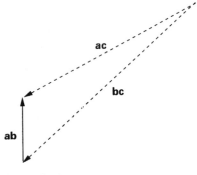

Figure E1.4

1.4.3   Polar–rectangular coordinate conversion

Once we have a resultant vector (or any vector) in rectangular coordinates it is often beneficial to describe the vector in polar coordinates (r, θ form). This conversion is carried out using basic rules of trigonometry.

   Consider any vector where $\mathbf{a} = \Delta x\mathbf{i} + \Delta y\mathbf{j}$ (Figure 1.6). The magnitude of vector **a** is often written as $|\mathbf{a}|$, in polar coordinates $|\mathbf{a}| = r$. The magnitude of **a** is defined by Pythagoras' theorem:

$$|\mathbf{a}| = r = \sqrt{\Delta x^2 + \Delta y^2} \tag{1.1}$$

The orientation of vector **a**, with respect to the x-axis, is often written as $\angle\mathbf{a}$, in polar coordinates $\angle\mathbf{a} = \theta$. The orientation of **a** is defined by basic trigonometry:

$$\angle\mathbf{a} = \theta = \tan^{-1}\left(\frac{\Delta y}{\Delta x}\right) \tag{1.2}$$

where $\tan^{-1}$ is the inverse tangent function available on all scientific calculators.

   Although equation (1.2) seems straightforward, it does contain a subtle trap. Consider the difference between 2/2 and −2/−2. The solution to both is 1, so your calculator will not know the difference. However there *is* a difference, it is illustrated in Figure 1.7

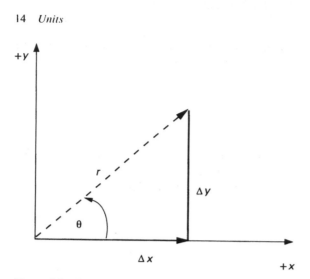

Figure 1.6   Any arbitrary vector

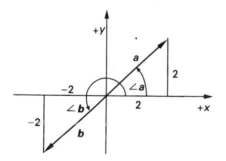

Figure 1.7   The difference between **a** = 2**i** + 2**j** and **b** = −2**i** − 2**j**

If you perform the calculation on your calculator, you will find that the angular orientation for both vectors is given as 45". This answer is fine for vector **a** but is incorrect for vector **b**. The reason for this is that the angle determined is the *included* angle, what we desire for **b** is the angle with respect to the *x*-axis. There is a simple rule to overcome this problem

If the value of the *x*-component (Δx) is less than zero add 180' to the answer.

We have seen that to add vectors together we need to know the *x* and *y* components of each vector. Since, by definition, a vector can point in any direction, we need to be able to reduce an arbitrary vector, given in polar coordinates, to its *x* and *y* components (the vectors used to describe any arbitrary vector in rectangular cordinates).

Consider the arbitrary vector **a**, given in polar coordinates (r, θ) as illustrated in Figure 1.6. Basic trigonometry is used to determine its *x* and *y* components. Firstly we always use the *x*-axis as our datum for measurement. Thus all angles must be measured from this axis. If this condition is satisfied then

$$\Delta x = r \cos \theta$$
$$\Delta y = r \sin \theta$$

(1.3)

where Δx is the *x*-component and Δy is the *y*-component of the arbitrary vector **a**.

If you own a scientific calculator (which is strongly recommended), you should learn how to use the 'polar to rectangular' and 'rectangular to polar' conversion functions.[3] This will save a great deal of time in later calculations because we shall be doing this conversion throughout the textbook.

In later studies you will become more conversant with vectors and, as your skills develop, you will find short cuts and tricks to reduce the time spent on building mathematical models. One of them is to change the datum axis. For now it is strongly recommended that you use the *x*-axis for a datum. When you feel more confident, the tricks will come naturally.

Example 1.5 *Add the following three vectors* **a**, **b** *and* **c** *(all measured with respect to the x-axis) to determine a resultant vector* **d**. *Give* **d** *in both rectangular and polar coordinates.*

$$\mathbf{a} = 120 \text{ at } 50°$$
$$\mathbf{b} = 56 \text{ at } 270°$$
$$\mathbf{c} = 75 \text{ at } 180°$$

SOLUTION
Firstly, we cannot add the three vectors together directly since they all point in different directions; we need to convert them to rectangular coordinates. As all angles are measured from the *x*-axis we can apply equations (1.3).

Table E1.5

Vector	x-component	y-component
a	$120 \cos (50) = 77.13$	$120 \sin (50) = 91.93$
b	$56 \cos (270) = 0$	$56 \sin (270) = -56$
c	$75 \cos (180) = -75$	$75 \sin (180) = 0$
d	2.13	35.93

From Table E1.5 we can write the vector **d** (in rectangular coordinates) as

$$\mathbf{d} = 2.13\mathbf{i} + 35.93\mathbf{j}$$

---

[3] This function is often signified by R–P and P–R, or by $\rightarrow r\theta$ and $\rightarrow xy$.

To convert **d** into polar coordinates we use equations (1.1) and (1.2) as follows:

$$|\mathbf{d}| = \sqrt{2.13^2 + 35.93^2} = 35.99$$

and

$$\angle\mathbf{d} = \tan^{-1}\left(\frac{35.93}{2.13}\right) = 86.6°$$

thus the resultant vector **d** = 35.99 at 86.6°. Graphically this can be represented as Figure E1.5.

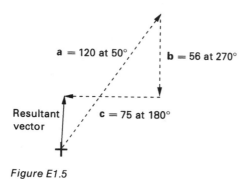

Figure E1.5

1.4.4 Vector multiplication

The multiplication of vectors is more complex. There are two methods of multiplying vectors, both termed *products* (because the result is a product of two vectors). The type of product depends on whether you want a scalar result or a vector result, hence they are termed *scalar (or dot) product* and *vector (or cross) product*. The actual mechanism of scalar and vector products is not necessary for this text but a study of their use reveals some useful short cuts for our analyses.

Consider two arbitrary vectors **a** and **b**. The scalar product (or dot product) is given by

$$\mathbf{a}.\mathbf{b} = ab\cos\theta$$

where a is the magnitude of **a**, b is the magnitude of **b** and θ is the angle of separation of **a** and **b**. It physically represents *the magnitude of* **a** *multiplied by the in-line magnitude of* **b**, as illustrated in Figure 1.8.

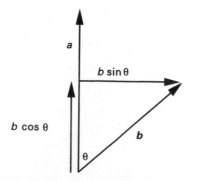

Figure 1.8 Graphical representation of a scalar (dot) product

When two vectors are aligned, such as when we use the *x*-axis only or *y*-axis only, the angle of separation θ is zero, hence $\cos \theta = 1$. In this situation the scalar product of **a** and **b** becomes

$$\mathbf{a}.\mathbf{b} = ab$$

The vector product (or cross product) of two vectors is given as

$$\mathbf{a} \times \mathbf{b} = ab \sin \theta$$

which is clearly zero when **a** and **b** are aligned (since $\sin(0) = 0$).

This fact will help us in later examples. To an engineer this means that, if we can reduce a problem to a model which we can analyse *one axis at a time*, all our vector equations can be reduced to scalar equations which are much easier to analyse.

1.5 Fundamental units

Those who are familiar with coloured pigments know that there are three primary colours, red, yellow and blue, from which all other colours can be mixed (four if you include black, which is an absence of colour). There are also fundamental units from which all others can be formed. Four units which are pertinent to this text are mass, length, time and temperature (see Table 1.1 for their full description).

As more quantities are met, such as speed and acceleration, we shall see how they are derived from these fundamental units.

1.6 Conversion of units

As briefly discussed in the introduction, the SI system of units is based on the metric system. Although engineers and scientists have been well schooled in this system since the 1960s, there are branches of industry which still deal with the Imperial system of units. Indeed it is still common to see some catalogues (although legislation is tending to reduce their number) which still use imperial units. Furthermore many of the imperial units have become part of the language and may take a long time to eradicate. So to complete this section, we shall explore the conversion of imperial to SI unit and vice versa.

Table 1.5 gives conversion factors for converting imperial units of length and mass to SI units and vice versa. As SI becomes adopted, this table will become less important.

Table 1.5 Conversion factors for mass and length

Unit	SI unit	Imperial unit
Mass	(lb × 0.454) kg	(kg × 2.205) lb
Length	(inch × 25.4) mm	(mm × 0.0394) inch
	(ft × 0.3048) m	(m × 3.281) ft

The unit for angle can also cause some difficulty because it may appear in three forms: degrees, radians and revolutions. In SI all angles should be in radians (later we shall see why this is important), but degrees are often used. To convert degrees and revolutions to radians remember that there are 2π radians in 1 revolution = 360°. This is set out in Table 1.6.

Table 1.6   Conversion factors for rad, deg and rev

Radian *(rad)*	Degree *(deg)*	Revolution *(rev)*
$(\deg \times \pi/180)$ rad	$(\text{rad} \times 180/\pi)$ deg	$(\text{rad}/2\pi)$ rev
$(\text{rev} \times 2\pi)$ rad	$(\text{rev} \times 360)$ deg	$(\deg/360)$ rev

The use of degree(°)minute(')second(") is not commonplace for the measurement of angles but the notation is easily converted, remembering that there are 60 minutes in 1 degree and 60 seconds in 1 minute (3600 seconds in 1 degree) . For example

$$1°30'45'' = 1 + (30/60) + (45/3600)$$
$$= 1.5125°$$

Example 1.6   *The steering-wheel of a saloon car turns through 3.5 revolutions when turning the wheels from hard right to hard left (lock to lock). Determine the rotation of the wheel in degrees and radians.*

SOLUTION

Angle in degrees: $3.5 \times 360 = 1260°$

Angle in radians: $3.5 \times 2\pi = 7\pi$ rad $= 21.99$ rad

(same as $1260\pi/180 = 21.99$ rad)

Summary

Units
SI (Système International) units are the preferred units.
SI units are specified in ISO 1000 (BS 5555) and ISO 35 (BS 5775).
Use the engineering units listed in Table 1.2.

Scalars and Vectors
A scalar requires only magnitude to describe it fully.
A vector has both magnitude and direction.
Vectors have a special form of mathematical manipulation called vector algebra.

The magnitude of a vector is given by

$$|\mathbf{a}| = r = \sqrt{\Delta x^2 + \Delta y^2}$$

The direction of a vector is given by

$$\angle \mathbf{a} = \theta = \tan^{-1}\left(\frac{\Delta y}{\Delta x}\right)$$

(if Δx is less than zero, add 180° to the result)
The x and y components of a vector are given by

$$\Delta x = r\cos\theta$$
$$\Delta y = r\sin\theta$$

if θ is measured with respect to the x-axis.

Problems

Coordinate systems

1.1 Sketch the following points on the x–y plane:

$$A = (1, 2)$$
$$B = (3, 4)$$
$$C = (-2, 2)$$

1.2 Determine the position of points A, B and C relative to the origin as illustrated in Figure P1.2.

1.3 Determine the position of point A relative to point B from Figure P1.2.

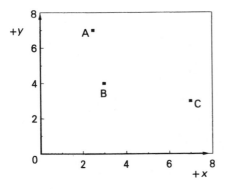

Figure P1.2

1.4 Determine the position of point A relative to point C from Figure P1.2.

1.5 Using the 3D axes x, y and z, sketch the following points:

$$A = (-1, 2, -1)$$
$$B = (2, 3, 2)$$
$$C = (3, -2, 3)$$

1.6 Using the data given in Problem 1.5, determine the position of points A, B and C relative to another point, D, where $D = (1, -2, 3)$.

Vector algebra

1.7   Using the data in Figure P1.2, determine vectors (in rectangular coordinates) which represents the displacement of points A, B and C from the origin.

1.8   Using the data in Figure P1.2, determine a vector (in rectangular coordinates) which describes the displacement of point C from point A.

1.9   Add the three vectors **a**, **b** and **c** detailed below and determine the resultant vector **d** in rectangular coordinates. Determine the answer both numerically and graphically.

$$\mathbf{a} = 45\mathbf{i} + 7\mathbf{j}$$
$$\mathbf{b} = 3\mathbf{i} - 15\mathbf{j}$$
$$\mathbf{c} = -20\mathbf{i} - 10\mathbf{j}$$

1.10   Using the data given in Problem 1.9 determine **a** − **b**, **b** − **a** and **c** − **a**.

1.11   Using the data given in Problem 1.9, determine **a** − **b** − **c**.

1.12   Convert vectors **a**, **b** and **c** in Problem 1.9 to polar coordinates.

1.13   Determine the x and y components of the following vectors and hence write them down in the form $\mathbf{i} + \mathbf{j}$:

$$\mathbf{ab} = 45 \text{ at } 76°$$
$$\mathbf{ac} = 65 \text{ at } 129°$$
$$\mathbf{ad} = 123 \text{ at } 225°$$

1.14   From the data given in Problem 1.13 determine the resultant vector (in both rectangular and polar coordinates) of the vector addition $\mathbf{e} = \mathbf{ab} + \mathbf{ac} + \mathbf{ad}$.

1.15   Convert the following vectors to polar coordinates:

$$\mathbf{a} = 70\mathbf{j}$$
$$\mathbf{b} = 45\mathbf{i} - 76\mathbf{j}$$
$$\mathbf{c}: x\text{-component} = 45, \ y\text{-component} = 37$$

Conversion of units

1.16   Rewrite the following numbers using scientific notation and prefix notation:

2563.45 m
0.045 m
0.000 002 35 m

1.17   Convert the following lengths:

14 inches (to m)
1 ft 3 in. (to mm)
0.003 inch (to m, mm and µm)

1.18   Convert the following values of mass:

24 lb (to kg)
14 st 7 lb (to kg)

1.19   Convert the following angular measurements:

4.56 rad (to deg)
45.7° (to rad)
3.25 revolutions (to deg and rad)
4.256 revolutions (to rad)

Chapter 2

Velocity and Acceleration

The study of bodies in motion is of prime importance to the engineer. There are two terms specifically related to the study of motion, *kinetics* and *kinematics*.

Kinetics is the study of unbalanced forces acting on a body, resulting in a study of its motion.

Kinematics is the study of the motion of a body irrespective of forces. It is an examination of the body's position in space with any associated velocities and accelerations.

In this chapter we shall be examining the motion of bodies which move in straight lines, often termed *rectilinear motion* or *linear motion*, and in circular arcs, often termed *circular* or *angular* motion. We shall also be examining any connection between these two forms of motion.

2.1 Linear motion

Objects that move in straight lines perform linear motion. Consider the motion of a football which is being passed diagonally across the field, and in a straight line, as illustrated in Figure 2.1.

The ball initially exists at position A; after some time interval Δt, the ball resides at position B. The displacement of the ball can be defined as the distance which separates B from A; since a direction is also required, displacement is a vector.

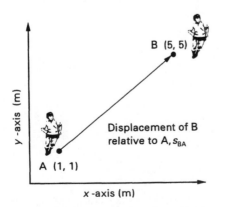

Figure 2.1 Linear motion of an object

Displacement is defined as the difference between two position vectors. In Figure 2.1 the displacement of B relative to A can be given in three forms: (i) in the x-direction, (ii) in the y-direction and (iii) along the line of motion. The displacements in these three cases are as follows:

Displacement in x: $x_{BA} = x_B - x_A = 5 - 1 = 4\,m$ (direction implicit)
Displacement in y: $y_{BA} = y_B - y_A = 5 - 2 = 3\,m$ (direction implicit)
Displacement: $s_{BA} = (5 - 1)i + (5 - 2)j = 4i + 3j$

the displacement s_{BA} may be written in polar form (see Chapter 1) as

$$|s_{BA}| = \sqrt{4^2 + 3^2} = 5\,m \quad \text{and} \quad \angle s_{BA} \tan^{-1}\left(\tfrac{3}{4}\right) = 36.9^{\circ}$$

or

$$s_{BA} = 5\,m \text{ at } 36.9^{\circ}$$

The scalar quantity, distance, is defined by the magnitude of s_{BA}.
   Velocity is defined as follows:[1]

Velocity (**u** or **v**) = rate of change of displacement

$$\approx \frac{\text{displacement}}{\Delta t} \approx \frac{(s_2 - s_1)}{(t_2 - t_1)} \tag{2.1}$$

Velocity is a vector quantity and its units are m/s or ms^{-1}. The scalar quantity, speed, is defined by the magnitude of equation (2.1).
   For motion in the x or y directions, equation (2.1) may be written

$$v_x \approx \frac{(x_2 - x_1)}{(t_2 - t_1)} \quad \text{and} \quad v_y \approx \frac{(y_2 - y_1)}{(t_2 - t_1)}$$

Strictly speaking, all vectors should be appropriately notated to distinguish them from other quantities (see Chapter 1). This is because the methods of mathematical manipulation are so different to those used for scalar quantities. However, when objects move in straight lines, it is only the magnitude of the velocity vector which changes. As such, they can be treated as scalar quantities in the context of developing engineering models. In these situations we have a choice of whether or not we use vector notation. In this text vectors will be denoted in bold.
   Consider now the scenario in which the time taken for the motion from A to B in Figure 2.1 was measured, the time at A being $t = 2\,s$ and at B 4 s. Note that the notation is such that at t_1 the object is at A, at t_2 it is at B. In this case, for the x and y directions

$$v_x = \frac{5 - 1}{2} = \frac{4}{2} = 2\,m/s \quad \text{and} \quad v_y = 3/2 = 1.5\,m/s$$

The overall velocity, i.e. the velocity of the object actually travelling from A to B, may be obtained in two ways: by using the x and y components of velocity

$$|v| = \sqrt{2^2 + 1.5^2} = 2.5\,m/s$$

or by examining the displacement of the body along the line of motion

$$v = 5/2 = 2.5\,m/s \text{ at } 36.9^{\circ}$$

note that the direction is identical to the displacement vector.

---

[1] $(s_2 - s_1)/(t_2 - t_1)$ is an approximation and is usually termed *average velocity*; the same applies to x and y motion and to acceleration.

The above definition for velocity is acceptable when velocity is constant over a time period. It is often the case, however, that velocity changes with time. We therefore need a more precise definition for velocity.

Figure 2.2 illustrates a situation where velocity is not constant, the actual displacement/time data is given in Table 2.1 (direction being constant). The data relates to an object accelerating from rest for a period of 7 s, displacement being measured at 1 s intervals.

Velocity may be defined by the gradient of the curve at any point (highlighted on Figure 2.2 as the tangent to the curve at $t = 5$ s). This is in fact very similar to a

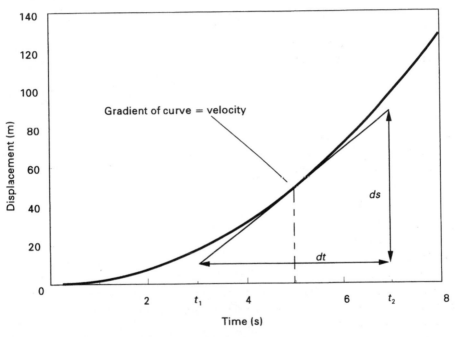

Figure 2.2   Variation of displacement with time

Table 2.1   Displacement–time data for Figure 2.2

Time, t *(s)*	Displacement s (m)
0	0
1	2
2	8
3	18
4	32
5	50
6	72
7	98

definition for the mathematical tool called *differentiation*, (a method for determining the gradient of any curve using standard mathematical methods).[2] For the purposes of linking displacement and velocity, using *differential calculus*, we can define velocity as

$$\mathbf{v}(t) = \frac{d(\mathbf{s}(t))}{dt} \left(\frac{d(x(t))}{dt} \text{ in the } x\text{-direction or } \frac{d(y(t))}{dt} \text{ in the } y\text{-direction} \right) \qquad (2.2)$$

Equation (2.2) differs from (2.1) in as much as it is exact, not an average. This is because $t_2 - t_1$ is considered to be very small (in fact approaching zero), thus Δt has changed to dt. Therefore the techniques of differentiation may be utilised. In equation (2.1), $\mathbf{s}(t)$ is an equation which represents displacement as a function of time t. Further relationships are formed when $x(t)$ is used for displacement in the x-direction and $y(t)$ for displacement in the y-direction (this extension will be explored further in Chapter 3).

Velocity may not be constant over a given period of time. The unit which links velocity with time is acceleration. Acceleration is defined as follows:

Acceleration (**a**) = rate of change of velocity

$$\approx \frac{\text{change in velocity}}{\Delta t} \approx \frac{\mathbf{v}_2 - \mathbf{v}_1}{t_2 - t_1} \qquad (2.3)$$

The units for acceleration are m/s^2 or m s^{-2}. Acceleration is also a vector, its scalar equivalent is also called acceleration. Using differential calculus, we can define acceleration as

$$\mathbf{a}(t) = \frac{d(\mathbf{v}(t))}{dt} \qquad (2.4)$$

where $\mathbf{v}(t)$ is an equation which represents velocity as a function of time.

Example 2.1   *An object moves from* A $= (5, 6)$ *to* B $= (3, 2)$ *in 4 s. Determine*

    (a) the object's velocity in the x-direction
    (b) the object's velocity in the y-direction
    (c) the overall velocity of the object

SOLUTION
Using the definition for velocity given by equation (2.1)

    (a)

$$\mathbf{v}_x = (x_2 - x_1)/(t_2 - t_1) = (3 - 5)/4 = -2/4 = -0.5 \,\text{m/s}$$

    (b)

$$\mathbf{v}_y = (y_2 - y_1)/(t_2 - t_1) = (2 - 6)/4 = -4/4 = -1.0 \,\text{m/s}$$

---

[2] Those students who are either unfamiliar or unsure of the techniques of differentiation are recommended to look at *Foundation Maths* by A Croft and R Davison, published by Longman. A table of common differentials is given in Appendix B. A knowledge of calculus is not essential to follow this text but is mandatory for further studies.

(c)   Using the vector algebra we met in Chapter 1

$$\mathbf{v} = -0.5\mathbf{i} - 1.0\mathbf{j}$$

or in polar form

$$|\mathbf{v}| = \sqrt{(-0.5)^2 + (-1)^2} = 1.12\,\text{m/s}$$

$$/\mathbf{v} = \tan^{-1}\left(\frac{-1.0}{-0.5}\right) = 63.43^{\circ}$$

but since $v_x < 0$ the angle is in fact $63.43 + 180 = 243.43^{\circ}$,

therefore $\mathbf{v} = 1.12\,\text{m/s}$ at 243.43°.

Example 2.2   *Using the data given in Table 2.1, draw graphs of displacement, velocity and acceleration versus time.*

SOLUTION

First we must recognise that, when estimating slopes of curves from experimental data, we must take care. We cannot use the simple estimate of average velocity given by equation (2.1). Instead we must look to more accurate methods of determing the slope of a line. To do this we use *numerical differentiation*. Appendix C illustrates two equations which may be used to determine the slope of a graph at any point using three points (not just two). Table E2.2 demonstrates how the equations are used to determine estimates of velocity and acceleration from the data given in Table 2.1

Using Table E2.2 we can construct graphs of displacement, velocity and acceleration versus time (Figure E2.2).

Table E2.2   Development of displacement, velocity and acceleration from Table 2.1

t (s)	s (m)	v (m/s)	a (m/s^2)
0	0	$\frac{1}{2}[(-3 \times 0) + (4 \times 2) - 8] = 0$	$\frac{1}{2}[(-3 \times 0) + (4 \times 4) - 8] = 4$
1	2	$\frac{1}{2}[(-3 \times 2) + (4 \times 8) - 18] = 4$	$\frac{1}{2}[(-3 \times 4) + (4 \times 8) - 12] = 4$
2	8	$\frac{1}{2}[(-3 \times 8) + (4 \times 18) - 32] = 8$	$\frac{1}{2}[(-3 \times 8) + (4 \times 12) - 16] = 4$
3	18	$\frac{1}{2}[(-3 \times 18) + (4 \times 32) - 50] = 12$	$\frac{1}{2}[(-3 \times 12) + (4 \times 16) - 20] = 4$
4	32	$\frac{1}{2}[(-3 \times 32) + (4 \times 50) - 72] = 16$	$\frac{1}{2}[(-3 \times 16) + (4 \times 20) - 24] = 4$
5	50	$\frac{1}{2}[(-3 \times 50) + (4 \times 72) - 98] = 20$	$\frac{1}{2}[(-3 \times 20) + (4 \times 24) - 28] = 4$
6	72	$\frac{1}{2}[(32) - (4 \times 50) + (3 \times 72)] = 24$	$\frac{1}{2}[(16) - (4 \times 20) + (3 \times 24)] = 4$
7	98	$\frac{1}{2}[(50) - (4 \times 72) + (3 \times 98)] = 28$	$\frac{1}{2}[(20) - (4 \times 24) + (3 \times 28)] = 4$

The method adopted to develop Table E2.2 seems powerful at first. However, it is an approximation at best and must be treated with caution. It is very important to consider the effect of the length of the time step in between each point; the greater the time period, the less accurate the estimate of slope becomes. As a general rule, the more rapid the change in slope, the more data points are required. But it is important to realise that numerical

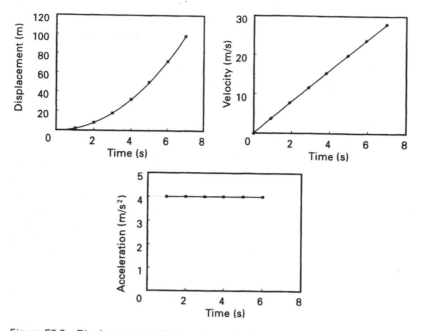

Figure E2.2   Displacement, velocity and acceleration curves for the data given in Table E2.2

differentiation is highly susceptible to noise – rapid changes of gradient – so numerical differentiation should normally be avoided. The lesson here is a practical one; if you are going to obtain experimental data, measure the data and integrate, i.e. acceleration to velocity to displacement.

Another trap is highlighted by the dotted line at $t = 0$ on the acceleration graph. Although it appears that the acceleration is constant at $4\,\text{m/s}^2$ from 0 to 7 s, we have to be sure this is what is actually happening. If the object were accelerating from rest, we would expect the acceleration and velocity to be zero at $t = 0$. In real life we would know the boundary conditions, but what we really need is more data between $t = 0$ and $t = 2\,\text{s}$.

Example 2.3   *A car travels forwards along a flat road with a constant speed of 144 km/h. The driver applies the brakes for 10 s, after which time the car is travelling at 36 km/h. Determine the acceleration of the vehicle over this time period.*

SOLUTION
The car's initial speed is

$$v_1 = 144 \times (1000/3600)$$
$$= 40\,\text{m/s}$$

The speed after 10 s is

$$v_2 = 36 \times (1000/3600)$$
$$= 10\,\text{m/s}$$

Since we know the car is travelling in the same direction all of the time, the direction is immaterial in this calculation. From equation (2.3), the average acceleration of the car is

$$a = \frac{v_2 - v_1}{t_2 - t_1}$$

$$= \frac{10 - 40}{10}$$

$$= -3 \, \text{m/s}^2$$

Note the $-$ (negative) sign for acceleration. One would normally use the term deceleration for this example but, although this is correct, one cannot always consider a negative acceleration as deceleration. The $-$ sign indicates that the acceleration vector is pointing backwards in relation to the car. In this case it has the effect of slowing the vehicle. In the case of a falling object, the acceleration is also negative (downwards), but the object's velocity increases with time. It is getting faster, but faster downwards! One must be careful not to confuse a negative acceleration with a deceleration, they need not be the same thing. In fact it is probably best to avoid the use of the term deceleration altogether!

So far we have examined the relationship between displacement, velocity and acceleration – differentiation. There is also a reverse link, *integration*, which is a mathematical tool for determining the area under any curve. Thus velocity attained (after a time period *t*) is the area under the acceleration–time graph, and displacement (after a time period *t*), is the area under the velocity–time graph (as demonstrated in Figure 2.3). To identify change in velocity or change in displacement use the time interval Δt, where $\Delta t = t_2 - t_1$; see equations (2.1) and (2.3)!

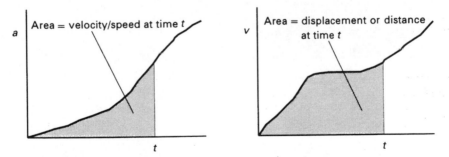

Figure 2.3   Area calculations for (a) velocity and (b) displacement

It is also possible to determine mathematical relationships:

$$v(t) = \int_{t_1}^{t_2} s(t)\, dt \quad \text{and} \quad a(t) = \int_{t_1}^{t_2} v(t)\, dt$$

As before, $s(t)$, $v(t)$ and $a(t)$ are equations describing distance travelled, speed and acceleration with time; similar equations apply to the vector equivalents.

Example 2.4   *The speed of a car was measured over several seconds and the data is given in Table E2.4; the car was travelling in a straight line. Using this information determine the distance travelled by the car over the period* $t = 0$ *to* $t = 8\,s$.

Table E2.4   Speed time relationship for a test vehicle

Time, t (s)	Speed, v (m/s)
0	4
1	6
2	8
3	10
4	12
5	12
6	12
7	10

SOLUTION

A graph of the speed relationship with time (velocity is inapproriate since we know that the car was travelling in the same direction all of the time) may be drawn, as illustrated in Figure E2.4.

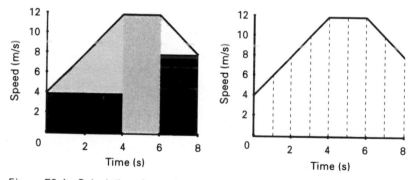

Figure E2.4   Calculation of speed: (a) by areas and (b) by Simpson's rule

The distance travelled over this period may be determined from the area under the graph. This may be obtained in two ways: (a) by geometry and (b) by Simpson's rule (Appendix C).

(a) $\text{Area} = (4 \times 4) + (0.5 \times 8 \times 4) + (12 \times 2) + (8 \times 2) + (0.5 \times 4 \times 2)$

$= 76\,\text{m}$

(b) $\text{Area} \approx \frac{1}{3}[4 + (4 \times 6) + (2 \times 8) + (4 \times 10) + (2 \times 12) + (4 \times 12) + (2 \times 12)$

$+ (4 \times 10) + 8]$

$= 228/3$

$= 76\,\text{m}$

Since velocity is a vector, the term *change in velocity* has another interpretation apart from just change in magnitude, it can also be change in direction. Figure 2.4 illustrates this for two vectors. An object has an initial velocity $\mathbf{v}_1 = 2\mathbf{i} + 2\mathbf{j}$ at time t_1 and a velocity $\mathbf{v}_2 = 2\mathbf{i}$ at time t_2; vector notation is mandatory in this case. The change in velocity is defined as

$$d\mathbf{v} = \mathbf{v}_2 - \mathbf{v}_1$$
$$= (2\mathbf{i}) - (2\mathbf{i} + 2\mathbf{j})$$
$$= -2\mathbf{j}$$

Since there is a change in velocity, there must be an acceleration

$$\mathbf{a} = d\mathbf{v}/dt \approx -2\mathbf{j}/(t_2 - t_1)$$

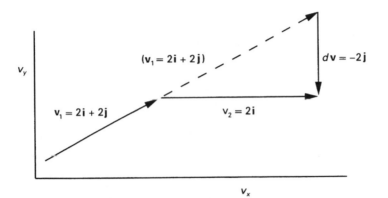

Figure 2.4 'Change in velocity' due to change in direction

2.2 Angular motion

In this section we shall be discussing the motion of objects moving in a circular arc about a single point of rotation. This is known as circular or angular motion. The definitions for angular displacement, angular velocity and angular acceleration are similar to those for linear motion. Consider an object moving in a circular arc as shown in Figure 2.5. In angular motion we normally work in *polar coordinates*, where

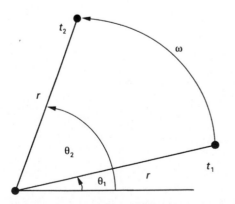

Figure 2.5 Rotation of a body about a centre

r is a radial position from an instantaneous centre and θ is an angular displacement from a known datum; θ is being measured in radians and an anticlockwise angular displacement is positive.

As with linear motion, angular displacement is defined as the angle which separates two points on a circular arc; in Figure 2.5 this is given by $\theta_2 - \theta_1$. We can define angular velocity in a similar fashion:

Angular velocity (ω) = rate of change of angular displacement

$$\approx \frac{\text{change in angular displacement}}{\text{time taken}} \approx \frac{\theta_2 - \theta_1}{t_2 - t_1} \qquad (2.5)$$

$$= \frac{d\theta}{dt}$$

The units for angular velocity are rad/s or rad s^{-1}.

Angular acceleration may be defined as follows:

Angular acceleration (α) = rate of change of angular velocity

$$\approx \frac{\text{change in angular velocity}}{\text{time taken}} \approx \frac{\omega_2 - \omega_1}{t_2 - t_1} \qquad (2.6)$$

$$= \frac{d\omega}{dt}$$

The units for angular acceleration are rad/s^2 or rad s^{-2}. We can apply to angular motion the same principles that we applied to linear motion. We can use a similar method to construct angular displacement–time graphs and we can derive similar relationships between angular velocity and angular acceleration.

Example 2.5   *On inspection, the central support of a fairground ride (Figure E2.5) was found to rotate 75 times in one minute; determine the angular velocity of the central support in rad/s. If the ride takes 2 minutes to reach its operating speed, determine the average angular acceleration.*

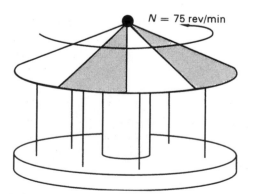

$N = 75\ \text{rev/min}$

Figure E2.5   Fairground ride

SOLUTION

If a body rotates 75 times in one minute, its rotational speed is given by $N = 75\,\text{rev/min}$.

To convert revolutions per minute (N) to angular velocity, remember that there are 2π radians in one revolution, and 60 seconds in one minute, therefore

$$\omega = N \times \frac{2\pi}{60} = 75 \times \frac{2\pi}{60} = 7.854\,\text{rad/s}$$

It is hard to define a sign for ω, since a wheel which rotates in one direction when looking at it from one side rotates in the other direction when looking from the other side! Hence with angular velocity it is not only important to give the sign but also the viewing position (looking from the front, etc.) but we still retain anticlockwise as positive rotation.

To solve the second part of the question we recall equation (2.6):

$$\alpha \approx \frac{\omega_2 - \omega_1}{t_2 - t_1}$$
$$= \frac{7.854 - 0}{120 - 0}$$
$$= 0.065\,\text{rad/s}^2$$

which, of course, assumes that the acceleration is constant throughout.

2.2.1 Simple harmonic motion

An application of the concept of a body rotating with angular velocity ω is *simple harmonic motion* (*SHM*). Simple harmonic motion is displayed by simple systems such as pendulums. It can also be displayed in complex systems, e.g. the vibration of buildings. If you consider a pendulum, its motion is effectively back and forth, never actually getting anywhere, which is much the same as going around a circular path. If you plot the displacement (x) of the pendulum from the midpoint against time, you get a graph as shown in Figure 2.6; this is a sine wave of form

$$x(t) = A \sin \omega t \tag{2.7}$$

where A is the amplitude of the sine wave, and ω is the frequency of the sine wave. The product ωt effectively yields angles, that is angles from 0 to $360°$ (0 to 2π radians). The frequency ω is identical to the angular velocity of an object moving in a circular path, thus it is sometimes called the *circular frequency*. Figure 2.6 demonstrates the link between a swinging pendulum and an object rotating with angular velocity ω.

You may wish to confirm this phenomenon with a simple experiment. Firstly attach a piece of tape (preferably coloured) to the rim of a bicycle wheel. Allow the wheel to rotate freely (by standing the bicycle upside down). Look at the wheel end on, you will see the tape move up and down in a rhythmic manner, performing simple harmonic motion.

The displacement–time curve illustrates that the object stops at the limits of travel and has greatest velocity when the amplitude is zero (at the midpoint). The time taken for the pendulum to swing one cycle (starting and finishing at the same point) is called the *periodic time* (t_p). It is easily found from $t_p = \omega/2\pi$. The velocity along the x-axis of the pendulum may be found from

$$v_x = \frac{dx(t)}{dt} = \frac{d(A \sin \omega t)}{dt} = A\omega \cos \omega t \tag{2.8}$$

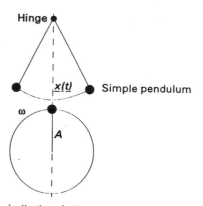

Hinge

x(t)   Simple pendulum

ω

A

Periodic time is the time taken for the pendulum to swing back and forth i.e. to perform one cycle

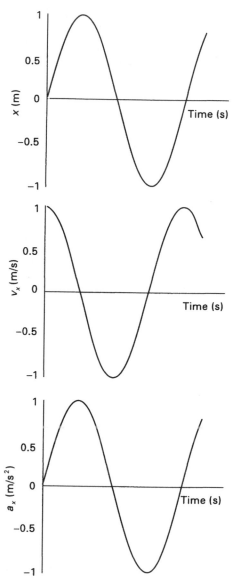

Figure 2.6   A body performing simple harmonic motion

and the acceleration from

$$a_x = \frac{dv_x(t)}{dt} = \frac{d(A\omega \cos \omega t)}{dt} = -A\omega^2 \sin \omega t \qquad (2.9)$$

A knowledge of basic trigonometry tells us that the maximum amplitude of a body moving with simple harmonic motion is A (the maximum of $\sin \omega t$ is 1), the maximum velocity is given by $\pm \omega A$, and the maximum acceleration is $\pm \omega^2 A$.

2.3   *Relationships between angular and linear motion*

Consider an object moving from position 1 to position 2 in a circular arc about a centre of rotation O, as shown exaggerated in Figure 2.7.

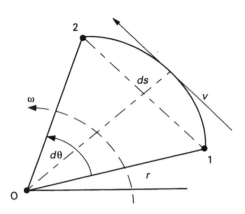

Figure 2.7   An object moving in a circular arc

The distance travelled over the time period dt (once again we assume dt, and hence $d\theta$ to be small) is given by the distance which separates position 1 and position 2. This is in fact equivalent to the length of the arc which separates 1 and 2,[3] which is given by

$$ds = r\,d\theta$$

The magnitude of the velocity vector is thus

$$|\mathbf{v}| = \frac{ds}{dt} = \frac{r\,d\theta}{dt} = r\frac{d\theta}{dt} = r\omega \quad \text{or} \quad v = \omega r \qquad (2.10)$$

Examination of Figure 2.7 demonstrates that we have determined a linear velocity for the object as it rotates, and that this linear velocity is actually a tangent to the curve at the point midway between positions 1 and 2. The sense of the vector is given by the direction of rotation. As $d\theta$ tends to zero, the midway point is almost identical to points 1 and 2, so the velocity is *always* a tangent to the arc, that is perpendicular to the radius of the arc. It is important to realise that this velocity is an instantaneous velocity and that the direction of the velocity vector depends on its position along the arc. If the object is travelling in a circle, one can imagine the velocity vector pointing up on one side and down on the other side, 180° around the periphery. If you are

---

[3] If $d\theta$ is very small (i.e. dt is very small), the error between the arc length and the cord length which joins positions 1 and 2 is also very small. The error is in fact $e = r\,(d\theta/2) - \sin(d\theta/2)$, which tends to zero as $d\theta$ tends to zero since $\sin(d\theta/2)$ tends to $d\theta/2$.

unsure of this, draw a large circle then draw tangents to the circle at $30''$ intervals; put arrowheads at the end of each tangent to indicate a vector; you will see how the velocity vector's direction changes with position.

Example 2.6   *A car is travelling horizontally with a forward speed of 75 km/h; the car is running on wheels of diameter 350 mm (forced to rotate about their centre). Determine the peripheral velocity and angular velocity of the wheels.*

SOLUTION

Two important concepts are to be learnt from this example; firstly *peripheral velocity.* Since linear velocity is given by $v = \omega r$, for a wheel or any circular object, any value of r can be selected from the centre of the wheel to its outermost edge, its *periphery.* Linear velocity is clearly zero at the centre, but at its periphery the linear velocity is a maximum; the peripheral velocity is therefore the linear velocity where r is a maximum (diameter/2). Figure E2.6, illustrates the second concept, that of bodies rolling *without slip.* If a wheel, barrel or any round object rolls without slip, the peripheral velocity of that object must be equal to the actual velocity of the object, but at the point of contact it is pointing in the opposite direction. In Figure E2.6 the peripheral velocity of the wheel must be identical to the forward velocity of the car (but in the opposite direction), that is

$$v = 75 \times 1000/3600$$
$$= 20.83 \, \text{m/s}$$

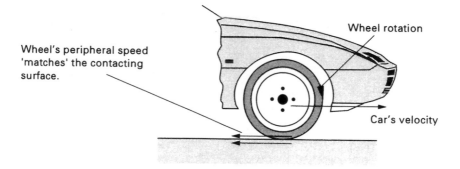

Wheel rotation

Wheel's peripheral speed 'matches' the contacting surface.

Car's velocity

Figure E2.6   A wheel rolling without slip

The angular velocity of the wheel comes from a knowledge of the peripheral velocity, i.e.

$$v = \omega r$$

which we can transpose to $\omega = v/r$, hence

$$\omega = 20.83/(0.350/2)$$
$$= 119.05 \, \text{rad/s} \quad (\text{or } N = 119.05 \times 60/2\pi = 1136.8 \, \text{rev/min})$$

Example 2.7 *A friction drive system is illustrated in Figure E2.7. The drive works on the basis that drive wheel A rolls without slip on driven wheel B; A has diameter 100 mm and B has diameter 250 mm. Determine the ratio of output speed to input speed for the system.*

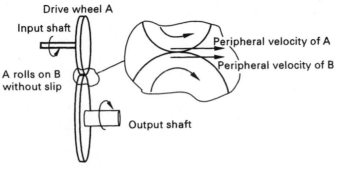

Figure E2.7 A geared drive system

SOLUTION
Since wheel A rotates without slip on wheel B, this means that their mutual peripheral velocities (at the point of contact) must be equal, as illustrated in the blow-up of the point of contact.

The peripheral velocity of A is given by

$$v_A = \omega_A r_A$$

the peripheral velocity of B is given by

$$v_B = -\omega_B r_B \quad \text{(note that } \omega_B \text{ is opposite in sense to } \omega_A\text{)}$$

since $v_A = v_B$, then

$$\omega_A r_A = -\omega_B r_B \quad \text{which leads to} \quad -\frac{\omega_B}{\omega_A} = \frac{r_A}{r_B}$$

This in turn leads to the following expression for gearing ratio:

$$\frac{\text{output speed}}{\text{input speed}} = \frac{-r_A}{r_B} = \frac{-D_A}{D_B}$$

which is in fact a general equation used for wheels in contact, and for gears in contact. Hence the gear ratio for this system is

$$\frac{\text{output speed}}{\text{input speed}} = -\frac{100}{250}$$
$$= -0.4$$

which is a reduction.

It was stated earlier that the body would prefer to move in a straight line, but what contains its motion so that it moves in an arc? An examination of the velocity vectors reveals a particular form of linear acceleration.

Figure 2.8 shows that, although there is no change in the magnitude of the linear velocity $v = \omega r$, there is a change in direction as the object moves from position 1 to position 2. The vector diagram demonstrates this and shows that the change in direction $d\theta$ is identical to the angular orientation of positions 1 and 2.

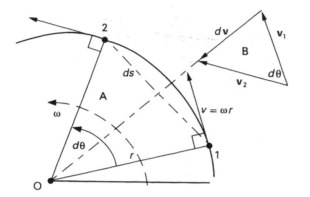

Figure 2.8   Change in velocity due to angular motion

We have already seen that $ds = rd\theta$. By the properties of similar triangles, the ratio of the sides of triangle A is equal to the ratio of the sides of triangle B, that is

$$\frac{ds}{r} = \frac{dv}{|\mathbf{v}|}$$

which on rearranging becomes

$$dv = |\mathbf{v}|\frac{ds}{r}$$

Since acceleration is defined as dv/dt, the linear acceleration of the body is given by

$$|\mathbf{a}_c| = \left|\frac{dv}{dt}\right| = \frac{|\mathbf{v}|}{r}\frac{ds}{dt} = \frac{|\mathbf{v}|^2}{r}$$

or

$$a_c = \omega^2 r \qquad\qquad (2.11)$$

and its direction is indicated by the vector $d\mathbf{v}$, which shows that the acceleration is always along the radius and towards the centre of rotation. This acceleration is called *centripetal acceleration* (hence the subscript $_c$) and it is the acceleration *required* to cause an object to travel in a circular path.

To fully understand this phenomenon, consider a model aircraft tied to a piece of string, as shown in Figure 2.9. It is clear that normally the aeroplane would fly in a straight line, yet when connected to the piece of string it flies in an arc. The string is actually pulling on the aircraft. The only direction it can pull in is towards the centre, thus the aeroplane is accelerating towards the centre *continuously*.

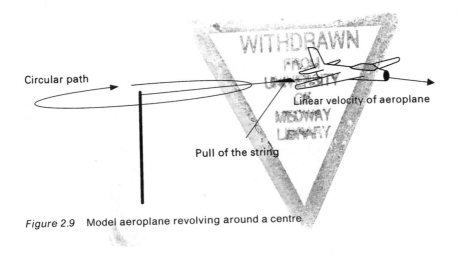

Circular path

Linear velocity of aeroplane

Pull of the string

Figure 2.9   Model aeroplane revolving around a centre

Example 2.8   *A motor cycle negotiates a bend in the road. If the bend has a radius equivalent to 400 m and the motor cycle is travelling at a constant speed of 95 km/h determine*

(a) *the angular velocity of the motor cycle*
(b) *the centripetal acceleration acting on the motor cycle*

SOLUTION

(a) First convert speed from km/h to m/s

$$v = 95 \times 1000/3600 = 26.39 \, \text{m/s}$$

Since we know that $v = \omega r$, we can transpose to

$$\omega = v/r$$
$$= 26.39/400$$
$$= 0.066 \, \text{rad/s}$$

(b) Centripetal acceleration is given by

$$a_c = \omega^2 r$$
$$= (0.066)^2 \times 400$$
$$= 1.74 \, \text{m/s}^2$$

2.4   Relative velocity and acceleration

We now have to consider the motion of bodies with respect to one another. It was highlighted earlier that displacement is a relative measurement and therefore depends on our datum. Both velocity and acceleration are relative, and they also depend on the choice of datum. Consider the following. Two people are saying farewell at a railway station. One person (person A) is on board the train, the other (person B) on the platform. The train starts to roll forward slowly, leaving the station. Person B walks forward with the train so as to keep this fond farewell going. To her the train, although moving, does not change position; the door and her devotee are still by her side. An observer would see the train, person A and person B moving forward with velocity **v**. Thus a measurement of velocity must always come with a statement of a datum. Normally one would assume a fixed point on the earth to be a datum. All too often one forgets that the earth moves in space, so the measurement of velocity with respect to ground is still relative.

Relative velocities are determined using vectors. Consider a section of motorway adjacent and parallel to a railway track (Figure 2.10), the cars are travelling with a constant speed of 120 km/h (relative to ground) but in opposite directions. The train is travelling with a constant speed of 160 km/h in the same direction as car A.

Figure 2.10 demonstrates the concept of relative velocity. The relative speed between car A and the train is $(160 - 120) = 40$ km/h. This is the same concept involved with overtaking - overtaking a stationary car takes less time than overtaking a car which is moving. Car B, however, is travelling in the opposite direction to the train, so the relative velocity between the car and the train is $(160 + 120) = 280$ km/h. To adhere to SI units the relative velocities may easily be converted to m/s.

The vector diagrams in Figure 2.10 demonstrate how these calculations are arrived at. Effectively, all we have carried out is a determination of the difference between two velocity vectors. The velocity of the train with respect to car A is signified by vector v_{TA}, which is determined from

$$v_{TA} = v_T - v_A$$
$$= 160\mathbf{j} - 120\mathbf{j}$$
$$= 40\mathbf{j}$$

As the vehicles are travelling along the same axis, this may be calculated simply by $v_{TA} = 160 - 120 = 40$ km/h.

The **j** simply signifies a vector in the *y*-direction. The velocity of the train is positive with respect to the car, so it will overtake the car and gradually the distance between the car and the train will increase in the positive direction (up the page). The velocity of the train with respect to car B is given by v_{TB}

$$v_{TB} = v_T - v_B$$
$$= 160\mathbf{j} - (-120\mathbf{j})$$
$$= 280\mathbf{j}$$

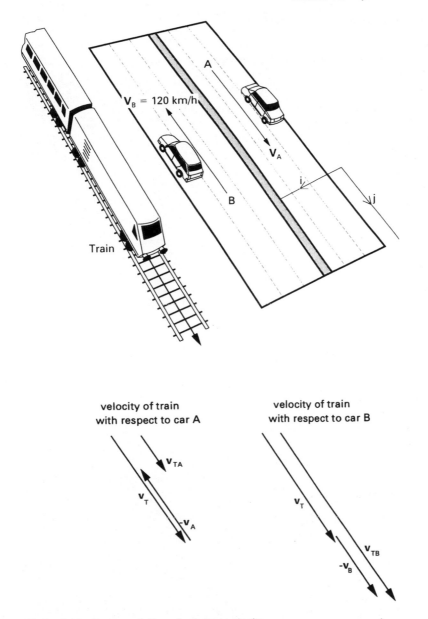

$V_B = 120$ km/h

Figure 2.10 Representation of relative velocity

Again, the velocity of the train relative to car B is positive, but is much greater. If only one axis is involved (i.e. direction **i** or **j**) then numerical addition or subtraction is acceptable; if the vectors are not parallel then vector addition and subtraction techniques must be applied (see Chapter 1). This principle is also applicable to displacement and acceleration.

***Example* 2.9**   *Two cars approach a junction at the same speed (72 km/h) but in different directions, as depicted in Figure E2.9. Determine the velocity of car A with respect to car O. Hence determine the velocity of car O with respect to car A.*

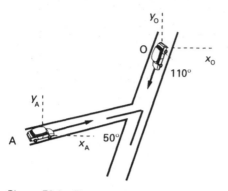

Figure E2.9   Two cars approaching a junction

SOLUTION

First convert speed in km/h to m/s

$$72 \times 1000/3600 = 20 \, \text{m/s}$$

The velocity of car O with respect to a horizontal datum (using axes x_0, y_0) is

20 m/s at $-110°$ or

$$v_O = 20 \cos(-110°)i + 20 \sin(-110°)j = -6.84i - 18.8j$$

Similarly for car A (using axes x_A, y_A)

$$v_A = 20 \cos(20°)i + 20 \sin(20°)j = 18.8i + 6.84j$$

Thus the velocity of car A with respect to car O is

$$v_{AO} = v_A - v_O = [18.8 - (-6.84)]i + [6.84 - (-18.8)]j = 25.64i + 25.64j$$

where the **i** and **j** components signify relative velocities in the x and y directions respectively.
   The velocity of car O with respect to car A is given by

$$v_{OA} = -v_{AO} = -25.64i - 25.64j$$

Summary

Linear motion

$\mathbf{v}$ = rate of change of displacement

$$= \frac{ds(t)}{dt} \approx \frac{s_2 - s_1}{t_2 - t_1} \, (\text{m/s})$$

or $\dfrac{dx(t)}{dt} \approx \dfrac{x_2 - x_1}{t_2 - t_1}$ in the x-direction

or $\dfrac{dy(t)}{dt} \approx \dfrac{y_2 - y_1}{t_2 - t_1}$ in the y-direction

$\mathbf{a}$ = rate of change of velocity $= \dfrac{d\mathbf{v}(t)}{dt} \approx \dfrac{\mathbf{v}_2 - \mathbf{v}_1}{t_2 - t_1} \, (\text{m/s}^2)$

Angular motion

$$\omega = \text{rate of change of angular displacement} = \frac{d\theta(t)}{dt} \approx \frac{\theta_2 - \theta_1}{t_2 - t_1} \ (\text{rad/s})$$

$$\alpha = \text{rate of change of angular velocity} = \frac{d\omega(t)}{dt} \approx \frac{\omega_2 - \omega_1}{t_2 - t_1} \ (\text{rad/s}^2)$$

The linear velocity at any point on a rotating body is given by

$$v = \omega r$$

It always acts perpendicular to the radius and in the sense of rotation.
    The centripetal acceleration of any point on a rotating body is given by

$$a_c = \omega^2 r = \frac{v^2}{r}$$

It always acts towards the centre of rotation.

Relative velocity
v_{AB} is the velocity of A with respect to B; in vector terms this may be written as
$v_{AB} = v_A - v_B$.

Problems

Linear motion

2.1   Define linear velocity. How are displacement and velocity determined from a knowledge of the relationship between acceleration and time?

2.2   Figure P2.2 illustrates three graphs of displacement versus time for an object moving linearly. In each case determine the velocity of the object (displacement x in metres, time t in seconds).

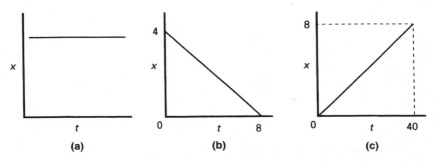

(a)   (b)   (c)

Figure P2.2

2.3   An aeroplane travels 5000 metres in two minutes, determine the speed of the aeroplane in m/s and km/h. What term is required for velocity to be known?

2.4   A ball on a pool table was found to move in a straight line from $x = 2\,m$ to $x = 0\,m$ in 2 seconds. What was the velocity of the ball?

2.5   At a test track, the acceleration of a car was examined by measuring the car's displacement from a start line (using a laser Doppler displacement probe) at 1 second intervals; the results from this experiment are given in Table P2.5. Draw graphs of displacement, velocity and acceleration versus time for the car and determine the average acceleration for the car over this period.

2.6   For the data given in Table P2.5 estimate the equations $s(t)$, $v(t)$ and $a(t)$ for the motion of the car.

Table P2.5

t (s)	1	2	3	4	5	6	7	8	9
s (m)	0.5	2	4.5	8	12.5	18	24.5	32	40.5

2.7   An object is dropped from a tall building. As the object drops, the velocity is measured at given intervals such that at $t = 4\,s$ $v = -39.24\,m/s$ and at $t = 8\,s$ $v = -78.48\,m/s$. Determine the acceleration of the object. What is the significance of the − sign?

2.8   At one instant in time, $t = 0$, a car is travelling along a flat road with a speed of 50 km/h. The driver presses the accelerator pedal for 10 s, after which time the car travels at 85 km/h. What is the acceleration of the car (in m/s^2)? You may assume constant acceleration.

2.9   Figure P2.9 illustrates a graph of velocity versus time for an object moving linearly. Determine the total distance travelled by the object over the period $t = 0$ to 30 s.

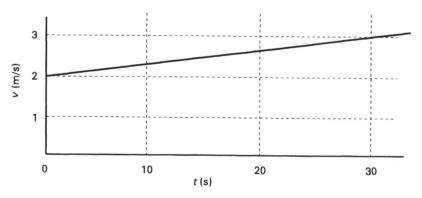

Figure P2.9

2.10   Figure P2.10 illustrates four graphs of velocity versus time; in each case determine the acceleration.

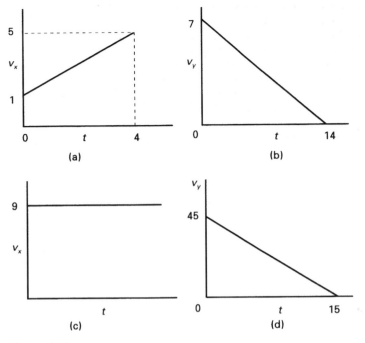

Figure P2.10

Angular motion

2.11 Define centripetal acceleration. What is significant about the centripetal acceleration of a body?

2.12 Show that the peripheral velocity of a wheel, of diameter D, rotating at N rev/min is given by $v = \pi N D/60$.

2.13 A link in a sewing-machine mechanism is oriented at $15°$ to the horizontal at $t = 0$ s; at $t = 0.01$ s the link has rotated to an angle of $176°$. Determine the angular velocity of the link.

2.14 A digger boom is illustrated in Figure P2.14. The whole body of the digger can rotate under the driver's control. When running at low speed the body rotates at 4 rev/min, at high speed the body rotates at 6 rev/min.

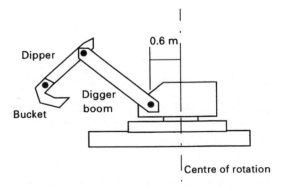

Figure P2.14

(a)   If the acceleration of the digger body is limited to 0.15 rad/s², determine
  (i)   the time taken to accelerate from rest to low speed
  (ii)   the time taken to accelerate from low speed to high speed
(b)   If the total length of the digger arm (boom, dipper and bucket) is 2.5 m when fully extended, determine the peripheral velocity of the arm at both low and high speeds.

2.15   If a disc of diameter 305 mm spins at a constant speed equivalent to 45 rev/min, determine
(a)   the angular velocity of the disc
(b)   the peripheral velocity of the disc
(c)   the centripetal acceleration acting on an object placed on the disc's periphery

2.16   A satellite attains a circular orbit above the earth at a height of 20 km. If the satellite orbits the Earth once every 1.5 hours, determine
(a)   the angular velocity of the satellite about the earth's centre
(b)   the linear velocity of the satellite
(c)   the centripetal acceleration acting on the satellite
The diameter of the earth is approximately 12 740 km. You may ignore the earth's movement in space.

2.17   A wheel of diameter 100 mm rotates at a constant speed of 1000 rev/min (clockwise). A second wheel runs in contact with the first wheel but is 150 mm diameter. Determine
(a)   the angular velocity of the first wheel
(b)   the peripheral velocity of the first wheel
(c)   the output speed of the second wheel

2.18   Show that the ratio of input speed to output speed for the gear system illustrated in Figure P2.18 is given by $N_1/N_3 = D_3/D_1$. Where N_1 and N_3 are the rotating speeds for the input gear and output gear respectively; D_1 and D_3 are their respective diameters.

Input gear, diameter D_1

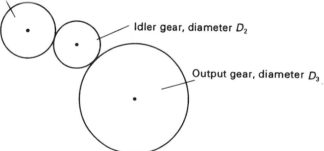

Idler gear, diameter D_2

Output gear, diameter D_3.

Figure P2.18

2.19   What is the function of the *idler gear* in the gearing system illustrated in Figure P2.18?

2.20   In an unfortunate accident a car's parking-brake has been left off and the car subsequently rolls down a hill. At the bottom of the hill the car has reached a speed of 45 km/h. If the car runs on wheels of 400 mm diameter determine the car's speed in m/s and the angular velocity of the wheels.

2.21   A pendulum swings back and forth exhibiting SHM with a periodic time of 2 s. If the amplitude of the swing is 0.025 m, determine equations for $x(t)$, $v(t)$ and $a(t)$ and subsequently state the maximum velocity and acceleration attained by the pendulum.

2.22 A flag-pole was observed to sway in the breeze and perform SHM where the displacement of the tip of the flag pole was given by $s(t) = 0.075 \sin 12.57t$. Determine the maximum velocity and maximum acceleration for the tip of the flag-pole.

Relative velocity

2.23 Two trains approach a junction, as illustrated in Figure P2.23. The speed of train A is 75 km/h. Determine the speed at which train B travels if they collide at the junction.

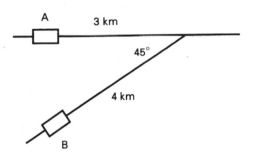

Figure P2.23

2.24 Aeroplane A leaves an airport 5 minutes after aeroplane B. The flight speed of A is a constant 190 km/h. The speed of B is 175 km/h. They are both travelling in the same direction.
(a) How long does it take for A to catch up with B?
(b) What is the relative velocity of B with respect to A?

2.25 If the aeroplanes in Problem 2.24 travel in opposite directions, what is the velocity of A with respect to B?

2.26 Figure P2.26 illustrates a simple mechanism in a sewing-machine. For the instant shown, point A on link OA has an absolute velocity of 5 m/s at 45" to the horizontal. Point B, on link AB, has an absolute velocity of 5 m/s at 90". Determine the relative velocity of A with respect to B.

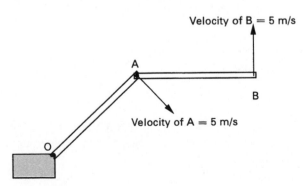

Figure P2.26

Chapter 3

Constant acceleration

In engineering there is a particularly useful assumption that one can make when determining the motion of objects – *constant acceleration*. This can mean one of three things: the object's speed is increasing linearly, the object's speed is decreasing linearly or the object's speed is not changing at all!

In this chapter we shall be developing three equations which can be used to model bodies whose acceleration is constant, in particular we shall be examining falling bodies in Section 3.2. The practical importance of assuming constant acceleration will be established in Section 3.3.

3.1 Development of the constant-acceleration equations

Firstly, what is meant by the term *constant acceleration*? Practically it means that $a = 0$, $a = c$ or $a = -c$ (where c is a constant). Using the $s(t)$ relationships for displacement versus time, seen in Chapter 2, we can establish two forms of $s(t)$ where acceleration is constant, as illustrated in Table 3.1.

Table 3.1 Polynomial functions of $s(t)$ where $a(t)$ is a constant.

Displacement/distance, $s(t)$	*Velocity/speed*, $v(t)$	*Acceleration*, $a(t)$
$s(t) = A + Bt$	$v(t) = B$	$a(t) = 0$
$s(t) = A + Bt + Ct^2$	$v(t) = B + 2Ct$	$a(t) = 2C$

The functions described in Table 3.1 are also demonstrated graphically in Figures 3.1 and 3.2, which illustrate that constant $a(t)$ requires the graph of acceleration versus time to be a horizontal, straight line.

Although the equations for $s(t)$, $v(t)$ and $a(t)$ describe the motion of an object exactly at any time, they can sometimes be cumbersome to use. To help us, we can develop some general equations relating to constant acceleration.

Figure 3.3 depicts the relationship between velocity (or speed) and time for an object which travels at constant velocity for 10 seconds then accelerates for a further 10 seconds.

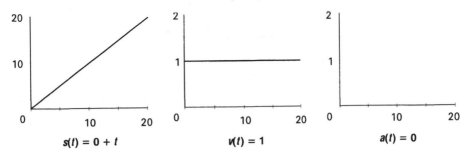

$$s(t) = 0 + t \qquad v(t) = 1 \qquad a(t) = 0$$

Figure 3.1 Linear displacement relationship

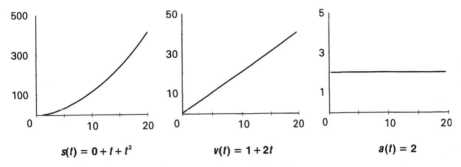

$$s(t) = 0 + t + t^2 \qquad v(t) = 1 + 2t \qquad a(t) = 2$$

Figure 3.2 Second-order displacement–time relationship

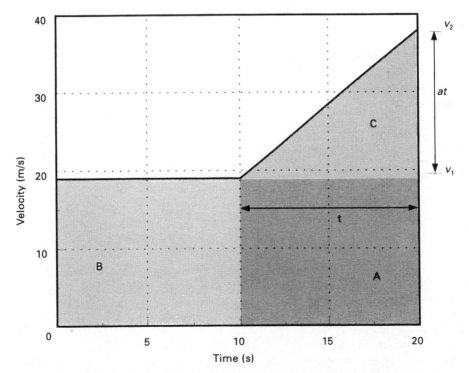

Figure 3.3 Velocity–time relationship for an object whose acceleration is constant

Over the period $t = 0$ to $t = 10$ s, the acceleration is constant at zero. Refering back to Chapter 2, the average acceleration for the object over the period $t = 10$ s to $t = 20$ s is given by

$$a = \frac{(v_2 - v_1)}{(t_2 - t_2)}$$
$$= \frac{(38 - 19)}{(20 - 10)}$$
$$= 1.9 \, \text{m/s}^2$$

The first constant-acceleration equation links the final velocity v_2, or speed, of an object with initial velocity v_1. Consider the following expression for average acceleration:

$$a = \frac{(v_2 - v_1)}{(t_2 - t_1)}$$

This can be rearranged to yield an equation for the final velocity v_2:

$$v_2 = v_1 + a(t_2 - t_1)$$

or

$$v_2 = v_1 + at \tag{3.1}$$

where t is the time period $(t_2 - t_1)$.

This is our first constant-acceleration equation. Note that equation (3.1) is written in its general scalar form but is equally applicable to vectors. Equation (3.1) is also demonstrated graphically on Figure 3.3. The second constant-acceleration equation relates to the displacement or the distance travelled over the time period t. Referring to Chapter 2 reminds us that the displacement of a body over a period $(t_2 - t_1)$, is the area under the velocity–time graph between t_2 and t_1. Consider again the time period $t = 10$ s to $t = 20$ s. The area under the graph for this period is Area A + Area C, where

Area A $= v_1 \times t$

and

Area C $= \frac{1}{2}(at \times t)$

Thus the displacement s attained over a time period t is given by

$$s = v_1 t + \frac{1}{2}at^2 \tag{3.2}$$

This is our second constant-acceleration equation. Equations (3.1) and (3.2) would normally be sufficient for most purposes, but a third relationship is required so that the equations are versatile in their use. Note that equation (3.1) uses v_1, v_2, a and t. Equation (3.2) uses s, v_1, a and t. The third equation produces a formula independent of t. From equation (3.1) we can determine a relationship for t, and substitute this into equation (3.2):

$$t = \frac{(v_2 - v_1)}{a} \tag{from 3.1}$$

therefore, by substitution into equation (3.2)

$$s = v_1 \left(\frac{v_2 - v_1}{a} \right) + \frac{1}{2} a \left(\frac{v_2 - v_1}{a} \right)^2$$

By multiplying out and using binomial expansion for the squared term, we get

$$s = \frac{v_1 v_2}{a} - \frac{v_1^2}{a} + \frac{1}{2a} \left(v_2^2 - 2v_2 v_2 + v_1^2 \right)$$

After simplification by subtraction, this becomes

$$s = \frac{v_2^2}{2a} - \frac{v_1^2}{2a}$$

which yields an expression for v_2 that does not involve t:

$$v_2^2 = v_1^2 + 2as \qquad (3.3)$$

Equations (3.1), (3.2) and (3.3) are the three constant-acceleration equations. They are so fundamental that they should all be memorised.

3.2 Gravitational acceleration

One of the most common forms of constant acceleration met by engineers, and indeed anyone, is acceleration due to gravity, or gravitational acceleration. Newton[1] established a law of gravitation which demonstrated that the acceleration achieved is inversely proportional to distance squared. When dealing with motion on or near to the earth's surface it is common to assume that this acceleration does not vary but is fixed at a nominal value of **9.81** m/s^2; see Appendix D.

Newton established that the acceleration is always directed towards the centre of the earth. And the acceleration always acts on the centre of mass of a body. Therefore gravitational acceleration always acts in the same direction, vertically downwards (as we normally view it).

Example 3.1 *A small child drops a stone down a well, which is known to be 12 metres deep. After a short time the child hears the plop as the stone hits the water.*

   (a) Determine how long it takes for the stone to drop 12 metres.
   (b) What is the stone's velocity when it hits the water?
   (c) Why is there a difference between the length of time the child measures and the actual time taken for the drop?

SOLUTION
Firstly let us sketch the problem by orienting the x-y axes such that y is vertical (Figure E3.1).

   Secondly let us examine the boundary conditions. We can examine the velocities and accelerations in the x and y directions at the instant when the object is released. Note that to simplify the problem we have oriented the y-axis with the vertical.

---

[1] Isaac Newton (1642–1727).

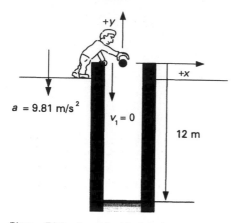

Figure E3.1   An object being dropped in a well

For the x-axis, initial velocity is zero and acceleration is zero. For the y-axis, initial velocity is zero (since it is being dropped) and acceleration is gravitational acceleration, 9.81 m/s² downwards (note that, for clarity, on Figure E3.1 the velocity is drawn as an arrow with one head and the acceleration as an arrow with two heads).

To solve part (a), we need an equation which links distance travelled with time and acceleration, we therefore use

$$s = v_1 t + \frac{1}{2} at^2$$

where we now know that

$s = -12$ m (because it is dropping downwards $- y$)

$v_1 = 0$

$a = -9.81$ m/s² (because it acts downwards)

therefore

$$-12 = (0 \times t) + (0.5 \times -9.81 \times t^2)$$

which when rearranged becomes

$$t = \sqrt{\frac{2 \times 12}{9.81}} = 1.56 \text{ s}$$

To solve part (b) we require an equation which relates final velocity to initial velocity, that is

$$v_2 = v_1 + at$$

from which

$$v_2 = 0 + (-9.81 \times 1.56) = -15.3 \text{ m/s}$$

(the $-$ sign tells us that the object is travelling downwards with respect to our chosen axes). This solution could have been achieved using equation (3.3).

Part (c) is slightly more intriguing. An initial solution could be that the child probably did not measure time to this accuracy. Secondly, one must not forget that the sound of the plop has to travel back up the well; this will take a finite length of time too!

3.3 Application of the constant-acceleration equations

In this section we shall be examining common applications of the constant-accleration equations. What matters is not the actual sequence of events but the methods of solution adopted.

3.3.1 One-dimensional motion

Firstly, what is meant by one-dimensional motion; as in Chapter 2 this means moving in a straight line where the direction never changes. Furthermore, since the direction is constant, we can use the constant-acceleration equations directly without concerning ourselves whether we are using vector or scalar quantities (since they will be the same).

Example 3.2 *A car is travelling along a straight stretch of road with an initial speed of 100 km/h. The driver presses the accelerator pedal which results in the car accelerating to a speed of 120 km/h in 12 seconds. Assuming that the car accelerates with constant acceleration, determine*

    (a) the acceleration of the car

    (b) the distance travelled over the period of acceleration

SOLUTION
For part (a) we need an equation which connects initial velocity, final velocity, time and acceleration; this is

$$v_2 = v_1 + at$$

or

$$a = (v_2 - v_1)/t$$

Before we can go any further, we need to convert speed in km/h to m/s; to do this we recognise that there are 1000 m in 1 km, and 3600 s in 1 h. Therefore

$$100 \text{ km/h} = 100 \times 1000/3600 = 27.78 \text{ m/s}$$
(a short cut is 100/3.6)

$$120 \text{ km/h} = 120/3.6 = 33.33 \text{ m/s}$$

hence

$$a = (33.33 - 27.78)/12$$
$$= 0.463 \text{ m/s}^2$$

For part (b) we require an equation which contains distance travelled; either equation (3.2) or (3.3) will do. We shall use equation (3.3):

$$v_2^2 = v_1^2 + 2as$$

After rearranging for s, we have

$$s = \left(\frac{v_2^2 - v_1^2}{2a}\right) = \left(\frac{33.33^2 - 27.78^2}{2 \times 0.463}\right) = 366.26 \text{ m}$$

Example 3.3   *A device for shooting tin cans vertically into the air (used in target ranges) is shown in Figure E3.3(a). If the can leaves the device with an initial velocity of 36 m/s, determine*

  (a)   the maximum height reached by the can
  (b)   the total time of flight

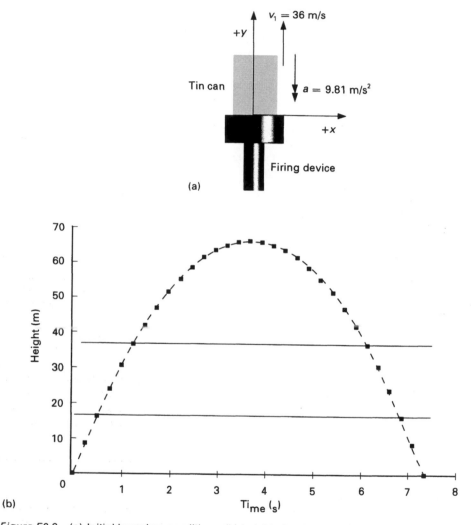

Figure E3.3   (a) Initial boundary conditions; (b) height of can versus time

SOLUTION

This is an interesting example because it highlights two facts, both concerned with objects that are fired upwards, reaching a zenith (maximum height) and subsequently dropping to the ground.

As in all engineering solutions, it is best to start with a sketch of the problem. Figure E3.3(a) indicates all the boundary conditions that we know. We have made some sweeping assumptions: (i) the can travels absolutely vertically, (ii) there are no side forces (e.g. wind) and (iii) that there are no effects caused by the air itself.

When an object is fired, or thrown, upwards into the air, it reaches a point where its motion changes from 'going up' to 'going down'. At this point the velocity has effectively changed from positive to negative. To do so, it must have gone through zero. For a small instant of time *the object actually stops.*

In essence, the solution to part (a) uses the following boundary conditions:

At $y = 0$: $v_1 = 36$ and $a = -9.81$

At $y = h_{max}$: $v_2 = 0$ and $a = -9.81$

where h_{max} is the maximum height attained by the can. Using

$$v_2^2 = v_1^2 + 2as$$

which on rearranging becomes

$$h_{max} \, s = \left(\frac{v_2^2 - v_1^2}{2a} \right) = \left(\frac{0^2 - 36^2}{2 \times -9.81} \right) = 66.06 \, m$$

The solution to part (b) is in itself simple, but there is a concept which is hidden within the solution; using

$$v_2 = v_1 + at$$

hence

$$t = (v_2 - v_1)/a$$
$$= (0 - 36)/(-9.81)$$
$$= 3.67 \, s$$

However, this is only for the object travelling upwards; the question asked for the total time of flight, so we must include the drop as well. This drop will take the same amount of time as the rise, 3.67 s. Thus the total time of flight is 7.34 s; note that we have assumed that the object starts and finishes in exactly the same place!

It is interesting to note that the tin can's motion on the way up is identical to its motion on the way down. You can demonstrate this by using the constant-acceleration equations to work out the can's positions at 0.1 s intervals, and plotting them on a graph of position versus time, as in Figure E3.3(b). The horizontal lines show that the can passes through each height twice, once on the way up and once on the way down. And if you move equal time steps each side of the zenith, you will find that the can is at the same height. The only difference is that, between the origin and the zenith, the can is moving upwards; from the zenith on, the can is moving downwards.

3.3.2 Angular motion

The use of the constant-acceleration equations for angular motion simply requires a substitution of the equivalent symbols for angular motion in place of the symbols for linear motion, as shown below (the linear equivalents are shown in parentheses):

$$\omega_2 = \omega_1 + \alpha t \quad (v_2 = v_1 + at) \tag{3.4}$$

$$\theta = \omega_1 t + \frac{1}{2}\alpha t^2 \quad \left(s = v_1 t + \frac{1}{2}at^2 \right) \tag{3.5}$$

$$\omega_2^2 = \omega_1^2 + 2\alpha\theta \quad (v_2^2 = v_1^2 + 2as) \tag{3.6}$$

Clearly the same sort of problems arise with angular motion as with linear motion.

Example 3.4 *The flywheel of an internal combustion engine, used to drive a pneumatic compressor, rotates at a nominal speed of 3600 rev/min. When the engine is switched off, it takes 2.5 seconds to come to a halt. If constant acceleration is assumed, determine*

> *(a) the average acceleration of the flywheel*
> *(b) the number of rotations the flywheel makes before it comes to a halt*

SOLUTION

To solve part (a) we use equation (3.4):

$$\omega_2 = \omega_1 + \alpha t$$

but the speed is in rev/min, so this needs to be converted to rad/s (see Chapter 2). The boundary conditions are $\omega_1 = 3600 \times 2\pi/60 = 377\,\text{rad/s}$, $\omega_2 = 0$, $t = 2.5\,\text{s}$. Therefore to determine the average acceleration

$$\alpha = (0 - 377)/2.5 = -150.8\,\text{rad/s}^2$$

For part (b) we use equation (3.5):

$$\theta = \omega_1 t + \frac{1}{2}\alpha t^2$$
$$\theta = (377 \times 2.5) + (0.5 \times -150.8 \times 2.5^2) = 471.25\,\text{rad}$$

which does not help us much since it is not in revolutions, thus we divide by 2π (since there are 2π radians in one revolution):

$$471.25/2\pi = 75\,\text{revolutions}$$

3.3.3 Two-dimensional motion

Many problems need to use the constant-acceleration equations in two dimensions. Examples of this form of motion are darts in flight, projectiles being fired from a plane and parcels being dropped from a plane. To understand the motion of objects under constant acceleration in two dimensions, we need to examine some fundamental concepts.

Consider a projectile being fired horizontally with an initial velocity of 75 m/s from the top of a cliff 250 m high. How far does the projectile travel horizontally before it hits the water?

Firstly we sketch the problem. The initial conditions are illustrated on Figure 3.4. Initially the projectile has a horizontal velocity of 75 m/s and its initial vertical velocity is zero. Also, the acceleration acting on the body can be examined along the horizontal and vertical axes. If we ignore the effects of air resistance and wind, the horizontal acceleration will be zero. The vertical acceleration is due to gravity and is therefore 9.81 m/s². To make the mathematics easy, we shall orient the vertical axis (up) with positive *y* and the horizontal axis with *x* (as shown on Figure 3.4) once again we treat the starting point as our datum for all measurements.

The usual question asked at this stage is, How can an object be moving both horizontally and vertically, at the same time, in such a manner? This question can be examined from two perspectives.

Firstly, consider that you are looking at the flight of the projectile from very high in the sky, so that it looks like a dot, as depicted in Figure 3.4. From your viewpoint

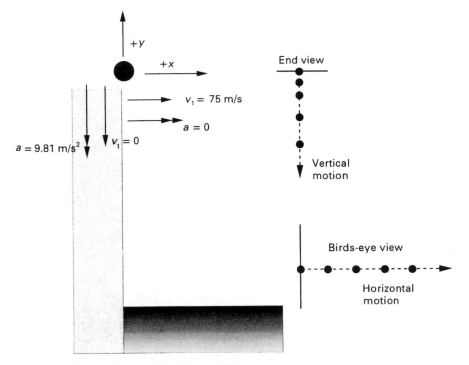

Figure 3.4 Projectile being fired from a cliff

the projectile will be moving along the ground, and you will not be able to see any upward or downward motion. In fact the only time you will realise that it is dropping will be when it hits the water. You will see the splash! This is motion along the x-axis and is represented by the bird's-eye view illustrated in Figure 3.4.

Now consider that you are on a sailing-vessel out at sea, and once again the projectile looks like a dot; you are viewing the projectile end-on so it is coming towards you. This time the motion towards you will not be visible, but the projectile will be clearly dropping. This is motion in the y-direction and is also illustrated in Figure 3.4.

Effectively, what we have done is to perform a simple piece of vector algebra. At any instant in time the object will have a single velocity vector. However, since we assume that the object only moves in two dimensions, i.e. planar motion, this single velocity vector can be split into two orthogonal components as shown in Chapters 1 and 2. This is a far more scientific answer, but maybe not so graphic! We select the axes of these components to be horizontal and vertical for ease of calculation.

To determine the horizontal distance covered, we need to determine the time of flight. To determine the time of flight, we need to calculate how long it takes for the projectile to drop the 250 m (the height of the cliff). We do the calculations first in the y-direction then in the x-direction (respectively oriented with the vertical and the horizontal).

y-axis

Using equation (3.2)

$$s = v_1 t + \frac{1}{2} a t^2$$

which we can rearrange for t, where $s = -250$, $v_1 = 0$ and $a = -9.81$

$$t = \sqrt{\frac{-250 \times 2}{-9.81}} = 7.14\,\text{s}$$

x-axis

Again we use

$$s = v_1 t + \frac{1}{2} a t^2$$

but with $v_1 = 75$, $a = 0$ and $t = 7.14$. This gives

$$s = 75 \times 7.14 + (0.5 \times 0 \times 7.14^2)$$
$$= 535.5\,\text{m}$$

Hidden within this example is another useful fact, which is the actual motion of any object in free flight. We can obtain a graph for the position of the object at any time interval, and this is depicted in Figure 3.5. The graph was obtained by calculating the distance travelled (in both x and y directions) by the projectile at 0.5 s intervals; you should attempt this for yourself. The motion prescribed is clearly a curve, it is a parabola; this is the motion of all objects which are in *free flight*, i.e. unpowered.

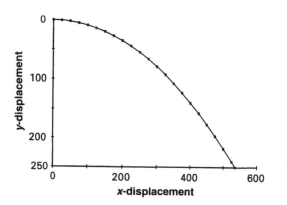

Figure 3.5   Motion of the projectile

This motion can be proved by considering a general case of motion in the *y*-direction with respect to motion in the *x*-direction. The time to move a given distance in the *x*-direction is given by

$$t = \frac{x}{v_1}$$

Substituting this into the expression for motion in the y-direction yields

$$y = -0.5 \times 9.81(x/v_1)^2 = -(4.905/v_1^2)x^2$$

which is of the form $y = Cx^2$, a general equation for a parabola, where $C = -(4.905/v_1^2)$. Hence the general motion for a projectile is always parabolic.

Example 3.5 *In the Olympic games, the javelin-throwing competition is one of the more spectacular events. Using the constant-acceleration equations, determine the best angle to throw the javelin so that it travels the furthest horizontal distance; you may assume that the athelete can throw the javelin with a maximum inital velocity of 20 m/s. In your calculations show that this angle is the projectile's optimum angle, for all initial velocities.*

SOLUTION
First let us sketch the problem at launch (shown in Figure E3.5), orienting the x-axis with the horizontal.

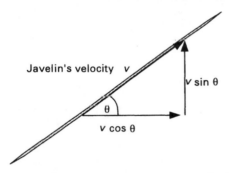

Figure E3.5 Initial conditions for a javelin throw

To solve this problem we need to identify two axes (x and y). We orient them with our normal horizontal and vertical axes for ease of calculation. This allows us to formulate boundary conditions; at the launch

y-axis; $v_1 = v \sin\theta$, $a = -9.81$ m/s^2
x-axis; $v_1 = v \cos\theta$, $a = 0$

at the top of the flight, we know that

y-axis; $v_2 = 0$

The total time of flight for the javelin is thus determined from the time it takes for the javelin to go up and come back down again, i.e. in the y-axis

$$v_2 = v_1 + at$$

or

$$0 = v\sin\theta + -9.81t$$

which can be rearranged to yield

$$t = v\sin\theta/9.81$$

But the total time of flight is $2t$, and therefore $2(v\sin\theta)/9.81$.

The horizontal distance travelled is therefore calculated using

$$s = v_1 t + \frac{1}{2} a t^2$$

which upon substitution of the initial conditions for the x-axis and the time of flight yields

$$x = (v \cos \theta) \cdot 2t$$

Now we can solve this problem in two ways. We can use a tabular method or solve it using classical methods. We shall attempt both solutions. To solve the problem using a table, we shall determine the time of flight and distance travelled at 7.5" increments, starting with horizontal (0") and finishing with vertical (90"). Table E3.5 illustrates the solution. The first column is the angle at which the javelin is thrown, the second and third columns are the horizontal and vertical components of velocity respectively, the fourth column is the time taken for the javelin to reach its maximum height and the final column is the total horizontal distance covered by the javelin.

Table E3.5   Motion of a javelin when thrown at different angles

θ	v *cos* θ	v *sin* θ	t	x
7.50	19.83	2.61	0.27	10.55
15.00	19.32	5.18	0.53	20.39
22.50	18.48	7.65	0.78	28.83
30.00	17.32	10.00	1.02	35.31
37.50	15.87	12.18	1.24	39.39
45.00	14.14	14.14	1.44	40.77
52.50	12.18	15.87	1.62	39.39
60.00	10.00	17.32	1.77	35.31
67.50	7.65	18.48	1.88	28.83
75.00	5.18	19.32	1.97	20.39
82.50	2.61	19.83	2.02	10.55
90.00	0.00	20.00	2.04	0.00

Table E3.5 highlights a problem, Why is the distance travelled zero when the javelin is thrown horizontally? Surely it must go somewhere! This has occured because we have assumed that the javelin is thrown from ground level, so throwing it horizontally has little effect (we shall see what happens when we start from a higher position in another example). The table does illustrate the fact that the optimum angle seems to be 45". Now we shall attempt to prove this analytically.

We have already determined an equation for the total time of flight, $t = 2(v \sin \theta)/9.81$. If we substitute this into $s = v_1 t + \frac{1}{2} a t^2$, we get an expression for the horizontal distance travelled, independent of t but dependent on the angle θ:

$$s = v_1 t + \frac{1}{2} a t^2$$
$$= (v \cos \theta) \cdot 2(v \sin \theta)/9.81$$
$$= (v^2/9.81) \cdot 2 \cos \theta \sin \theta$$

Using the double-angle formulae given in Appendix A, this may be simplified to

$$x = (v^2/9.81) \sin 2\theta$$

For this to be a maximum, sin 2θ must be as large as possible, i.e. 1; hence 2θ = 90", and the optimum angle θ is 45". It is the optimum angle for all velocities; so if you wish to throw something over a long distance, it is best to throw it at an angle of 45".

Example 3.6 *A new design of fire-engine uses a fixed hose on top of the engine which can emit a water jet at any desired angle with an exit velocity of 30 m/s. Determine the maximum range of the hose if the end of the hose is 2.3 m above ground level.*

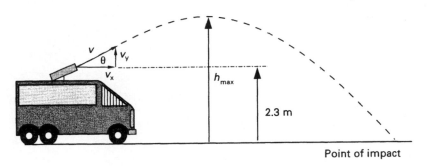

Figure E3.6 Sketch of the water jet's path

SOLUTION
To solve this problem we use the fact, already determined in Example 3.5, that the optimum angle is 45". However, in this case, to determine the total time of flight, we must calculate the time for the water jet to reach its maximum height then determine the time taken to reach the ground, since they will be different. Furthermore, we can use the constant-acceleration equations for water jets as well as projectiles because a water droplet, in terms of this basic model, is no different to any other form of projectile we care to imagine.

As in Example 3.5, we take horizontal and vertical components of the exit velocity (Figure E3.6). The time taken for the water jet to reach the maximum height is

$$t = \frac{v \sin \theta}{9.81} = \frac{30 \sin 45"}{9.81}$$
$$= 2.16 \, s$$

To determine the time taken for the water jet to drop from the zenith to ground, we must first determine the maximum height attained relative to the end of the hose. Using equation (3.2), we can write an equation which gives us this value:

$$y = (2.16 \times 30 \sin 45") + (0.5 \times -9.81 \times 2.16^2)$$
$$= 22.93 \, m$$

The total height through which the jet drops is thus $22.93 + 2.3 = 25.23$ m. We can now determine the time taken for the jet to drop this distance using equation (3.2) again:

$$-25.23 = (0 \times t) + (0.5 \times -9.81 \times t^2)$$

hence

$$t = \sqrt{\frac{25.23 \times 2}{9.81}} = 2.27 \, s$$

Which is slightly longer than the time taken to reach the zenith, as expected. The total time of flight is therefore $2.16 + 2.27 = 4.43\,\text{s}$. We can now determine the range of the water jet from the horizontal distance covered in this time, a similar solution to that given in Example 3.5. Using equation (3.2)

$$s = v_1 t + \frac{1}{2} a t^2$$

$$= (30 \cos 45^\circ \times 4.43) + \frac{1}{2}(0 \times t^2)$$

$$= 93.96\,\text{m}$$

Summary

Linear motion

$$v_2 = v_1 + at$$
$$s = v_1 t + \frac{1}{2} a t^2$$
$$v_2^2 = v_1^2 + 2as$$

where

v_1 = initial velocity
v_2 = final velocity
s = displacement or distance travelled
a = acceleration
t = time

Angular motion

$$\omega_2 = \omega_1 + \alpha t$$
$$\theta = \omega_1 t + \frac{1}{2} \alpha t^2$$
$$\omega_2^2 = \omega_1^2 + 2\alpha\theta$$

where the symbols represent the angular counterparts of the linear variables.

For objects which are in free flight, and subject to gravitational acceleration, two important boundary conditions can be applied:

(i)   If an object is fired, or thrown, so that it has a vertical component of velocity, the object will reach a maximum height before it begins to fall back to ground. At this point the vertical component of velocity is zero.

(ii) The optimum angle at which to fire, or throw, an object such that it travels the furthest horizontal distance is 45° to the horizontal.

Problems

One-dimensional motion

3.1   A shooting-range target is fired vertically and reaches a height of 185 m. Determine the initial velocity of the target.

3.2   For the data given in Problem 3.1, determine
   (a) the time taken to reach the maximum height
   (b) the time taken to reach 75 m

3.3   A 15 kg stone is dropped from the roof of a 25 m high building. Determine the time taken for the stone to reach the ground and the maximum velocity attained.

3.4   A jet of water is fired vertically upwards from the end of a hose. The end of the hose is 3 m above ground level. If the exit velocity of the water is 25 m/s, determine the height of the fountain produced and the time taken for a particle of water to reach the ground.

3.5   A car accelerates from rest. After 30 s the car's speed is 140 km/h. Assuming constant acceleration, determine
   (a) the car's acceleration
   (b) the distance travelled during this period

3.6   An ejector system fires an object vertically with an initial velocity of 80 m/s. Determine the maximum height attained by the object and the time taken to achieve it.

3.7   A car's speed is intially 150 km/h. The driver applies constant braking for 2 s, after which the car's speed is 100 km/h. Determine the acceleration of the car and the distance travelled during this period.

Angular motion

3.8   A flywheel rotates at 300 rad/s. A brake is applied for 10 s, which slows the flywheel's speed to 125 rad/s. Determine the acceleration of the flywheel. Also determine the number of revolutions performed by the flywheel during this period.

3.9   A flywheel is allowed to rotate at a constant speed of 50 rev/min for a period of 15 s. It is subsequently accelerated at 15 rad/s^2 until its rotational speed is 3000 rev/min. Determine the acceleration time to achieve this final speed and the total number of revolutions performed (including the period of constant speed).

3.10   A car is supported on wheels of 300 mm diameter. The car accelerates from rest to 125 km/h in 25 s. Determine the number of revolutions of each wheel during this period.

Two-dimensional motion

3.11   An object is ejected from the window of a tall building. The window is 50 m from the ground and the intial velocity of the object is 15 m/s horizontally.
   (a) Sketch the motion of the object including all boundary conditions.
   (b) Determine the time taken for the object to hit the ground.
   (c) Determine the horizontal distance covered by the object over the period determined in part (b).

3.12   An aeroplane flies horizontally at a height of 300 m with a forward speed of 200 km/h. If at some instant a parcel is dropped from the plane, determine
   (a) the time of drop
   (b) the horizontal distance covered by the object

3.13   A dart player throws a dart at a target which is 3 m away. The dart is intially thrown at the same level as the target and also horizontally. If the dart hits the backboard 0.3 m below the target determine
   (a) the time of flight
   (b) the inital velocity of the dart

3.14   Figure P3.14 illustrates a ski-jumper leaping from the end of the 'hill'. If the initial velocity of the person is 62 km/h horizontally determine the distance *s* travelled down the slope.

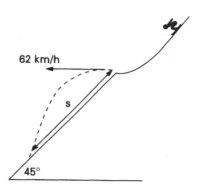

62 km/h

s

45°

Figure P3.14

3.15   A firework rocket accelerates at a constant 11 m/s² for a period of 1.5 s. If the rocket is fired vertically, determine
   (a)   the height achieved after the 1.5 s acceleration period
   (b)   the maximum height attained by the rocket
   (c)   the time taken to reach maximum height
   You may assume there is no sidewind.

3.16   Another rocket, similar to that in Problem 3.15, is fired at an angle of 60° to the horizontal. Recalculate parts (a), (b) and (c) for this condition.

3.17   The hose of a fire-engine expels water with a velocity of 25 m/s. The end of the hose is fixed at 2.6 m above ground level but can alter its attack angle from 0 to 85°, measured to the horizontal. Determine the maximum range of the water jet.

3.18   For the fire-hose in Problem 3.17 determine the attack angle required if the water jet is to *drop* onto a fire displaced from the hose 15 m horizontally and 25 m vertically.

Chapter 4

Newton's laws, momentum and force

All engineering mechanics is based on *Newtonian mechanics.*[1] Although more sophisticated theorems have been developed over the latter part of this century, the mainstays of engineering analysis are the theories developed by Newton. In this section we shall be examining momentum, impulse and force. We shall be examining collisions, their mechanics and how they can be modelled. We shall also be examining why objects move, and why they do not.

In particular we shall be examining Newton's laws of motion and their application to engineering mechanics.

4.1 Momentum and impulse

4.1.1 Linear momentum

Consider a body of mass m, travelling with linear velocity v (as illustrated in Figure 4.1).

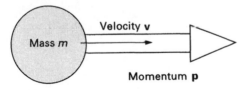

Figure 4.1 Mass moving with velocity **v**

The body is said to have linear momentum, where linear momentum is defined as the product of mass and velocity:

$$\mathbf{p} = m\mathbf{v} \tag{4.1}$$

Since velocity is a vector, linear momentum is also a vector; it has the symbol **p** and its units are kg m/s, later we shall see that N s can also be used.

---

[1] Isaac Newton (1642–1727) developed theories of motion which were presented in his famous publication *Principia* (1687).

Example 4.1 *A locomotive of mass 150 tonnes travels horizontally with a velocity of 75 km/h. Determine the linear momentum of the locomotive.*

SOLUTION
Using equation (4.1)

$\mathbf{p} = m\mathbf{v}$

Converting the speed of the train to SI units

$v = 75/3.6 = 20.833\,\text{m/s}$ (in the direction of the train)
$\mathbf{p} \overset{*}{=} (150 \times 1000) \times 20.833$

$= 3.125 \times 10^6 \,\text{kg m/s}$ (the direction is the direction of the locomotive)

4.1.2   Angular momentum

Consider a body of mass m rotating about a point with angular velocity ω as shown in Figure 4.2. The instantaneous linear momentum of the mass is still $\mathbf{p} = m\mathbf{v}$, but the body also rotates about O. The product of linear momentum and radius r is called the moment of momentum, or *angular momentum*.

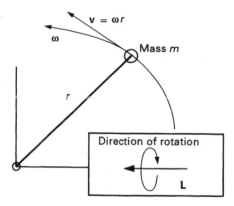

Figure 4.2   Derivation of angular momentum for a body rotating about a point

The magnitude of angular momentum is defined as

$L = pr$
$= mvr$

From Chapter 2 we know that $v = \omega r$, hence

$L = (m\omega r)r$
$= m\omega r^2$
$= I\omega$

or (in vector form)

$L = I\omega$ \hfill (4.2)

where I is called the moment of inertia of the body; it is the inertial property of a *rotating body*, much the same as mass is the inertial property of a body in linear motion. In this case, the inertia of a point mass about a centre of rotation is $I = mr^2$, but this is not true for all bodies. The determination of I for any shape is beyond the scope of this text. We shall be meeting moment of inertia as a number only, but note that it is always referenced to the axis of rotation.

Angular momentum has the symbol **L** and its units are $kg\,m^2/s$ (or Nms), it is also a vector quantity whose sense is determined by the right-hand screw rule. Figure 4.3 illustrates the right-hand screw rule; imagine you are opening or closing a screw-top on a drinks bottle and that will give the sense of the vector.

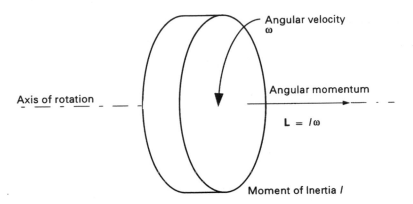

Figure 4.3 Flywheel rotating with angular velocity ω

All rotating bodies have angular momentum be they windmills, flywheels or ice-skaters.

Example 4.2 *A car of mass 1.5 tonnes is travelling with a speed of 115 km/h. The car runs on wheels of diameter 300 mm and whose moment of inertia is 0.85 kg m². Determine*

    (a) *the linear momentum of the car*
    (b) *the angular momentum of each wheel*

SOLUTION
    (a) First convert to standard units

$$1.5 \text{ tonnes} = 1.5 \times 1000 = 1500 \, kg$$
$$115 \, km/h = 115/3.6 = 31.94 \, m/s$$

Hence

$$p = 1500 \times 31.94 = 47910 = 47.91 \times 10^3 \, kg\,m/s$$

    (b) Recalling the solution to Example 2.6, the angular velocity of the wheels can be determined since the peripheral velocity of each wheel must be the same as the linear velocity of the car. Further we recall that

$$v = \omega r$$

or on rearranging

$$\omega = v/r$$

the radius of each wheel is $300/2 = 150\,mm = 0.15\,m$, therefore

$$\omega = 31.94/0.15 = 212.93\,rad/s$$

hence the angular momentum of each wheel is

$$\mathbf{L} = I\boldsymbol{\omega} = 0.85 \times 212.93 = 180.99\,kg\,m^2/s$$

We do not know direction, so only the magnitudes of linear and angular momentum have been determined.

Example 4.3 *A flywheel of moment of inertia 2.5 kg m² rotates at 3600 rev/min, as illustrated in Figure E4.3. Determine the angular momentum of the flywheel and sketch the sense of the momentum vector.*

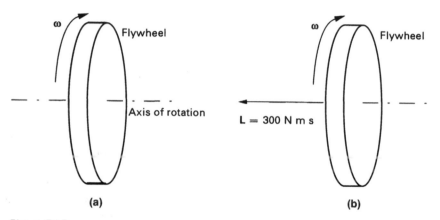

(a)   **(b)**

Figure E4.3

SOLUTION

First we need to convert rotating speed of the flywheel to SI, recalling that

$$\omega = 2\pi N/60 = 2\pi\,3600/60 = 120\pi\,rad/s$$

Recalling that angular momentum is defined by

$$\mathbf{L} = I\boldsymbol{\omega} = 2.5 \times 120\pi = 300\pi\,kg\,m^2/s$$

the sense of $\mathbf{L}$ is given by the right-hand screw rule, which from Figure 4.2 is from right to left, as shown in Figure E4.3(b).

4.1.3   Impulse

A transfer of momentum often occurs when two bodies collide or when one impinges on the other. This transfer of momentum forms the basis of Newtonian mechanics, and its value is defined by *impulse*.

Consider a body of mass m, which at one instant in time has a velocity $\mathbf{v}_1$. After an event, such as a collision, the body's velocity is $\mathbf{v}_2$ (as illustrated in Figure 4.4). The body's momentum before the event was

$$\mathbf{p}_1 = m\mathbf{v}_1$$

the momentum after the event was

$$\mathbf{p}_2 = m\mathbf{v}_2$$

The change in the body's momentum is therefore

$$\Delta \mathbf{p} = \mathbf{p}_2 - \mathbf{p}_1 = m\mathbf{v}_2 - m\mathbf{v}_1 = m(\mathbf{v}_2 - \mathbf{v}_1)$$

Figure 4.4 Change in momentum of a body due to an event

This change in momentum of a body is termed impulse. If the change is positive the momentum of the body has increased, because something has acted on it. If the change is negative the momentum of the body has decreased, probably because the body has acted on something else. So impulse is a measure of *effect*, the effect of an action, collison or event. Later we shall see that this term, be it impulse or change in momentum, is very important.

Note that the mass of a body can change (consider a rocket whose fuel is destroyed and emitted to produce thrust), so to be complete we should define impulse by

$$\Delta \mathbf{p} = \mathbf{p}_2 - \mathbf{p}_1 = m_2\mathbf{v}_2 - m_1\mathbf{v}_1 \qquad (4.3)$$

The unit of impulse is the same as momentum; it is a vector quantity.

Example 4.4 *A car of mass 1.3 tonnes travelling at 55 km/h collides with a stationary lorry. After the collision the car is stationary. Determine the impulse which acted on the car and state whether momentum was lost or gained.*

SOLUTION
First convert all non-SI units to SI

$$v_1 = 55/3.6 = 15.28 \, \text{m/s}$$
$$m = 1.3 \times 1000 = 1300 \, \text{kg}$$

Note that momentum can only be described by its magnitude (scalar) because we do not know direction of travel. After the impact the vehicle is stationary; thus $v_2 = 0$, hence

$$\Delta p = 1300 \times (0 - 15.28) = -19.861 \times 10^3 \, \text{kg m/s}$$

Since the sign is negative, this suggests that momentum was lost by the car.

4.2   Impacts

We have seen, in Section 4.1.3, that an event such as a collision can have an effect on the momentum of a body. Another criterion to consider is *how* the bodies collide – whether the collision was *plastic* or *elastic*.

Consider two bodies which are about to collide, one we shall call body A whose velocity is v_a, a second body we shall call B with velocity v_b (as shown in Figure 4.5). When the objects first collide they both deform at the point of contact. After a very short period of time the bodies begin to 'push back' against one another (similar to the 'push' one feels when squeezing a football). During the impact some momentum from one body may be transmitted to the other. Also the bodies may alter shape slightly as a result of the collision. The action of deformation associated with the transmission of momentum has a price – a loss of energy.[2] The mechanisms which take place during a 'real-life' impact are very complex, so the model below should be treated, at best, as an estimate.

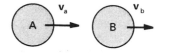 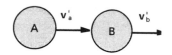

Before impact   During impact   After impact

Figure 4.5   Two bodies during impact

The impact may be modelled using

$$(v_b' - v_a') = e\,(v_a - v_b) \tag{4.4}$$

where

$$(v_b' - v_a') = \text{the } \textit{normal } \text{velocity of separation after impact}$$
$$(v_a - v_b) = \text{the } \textit{normal } \text{velocity of approach prior to impact}$$
$$e = \text{the } \textit{coefficient of restitution}$$

There are three types of collision that can occur

   (a)   *Perfectly elastic*, $e = 1$: the collision is perfect and no energy is lost due to the collision.

   (b)   *Perfectly plastic*, $e = 0$: both bodies become embedded within one another so that they act as one body thereafter. A great deal of permanent deformation occurs and the loss of energy in this type of collision is the greatest.

---

[2] Energy is defined in Chapter 5.

(c)   *Real impact*, $0 < e < 1$: in general the coefficient of restitution lies somewhere between 0 and 1. Some energy is lost while transmitting momentum from one body to the other (the reason why the bounces of a bouncing ball decrease with each impact).

Note that equation (4.4) stipulates *normal velocity* (v_n). The term *normal* is common in engineering and means 'perpendicular to a surface' at the point of contact. This is not a problem when the two bodies collide along the *line of contact*, producing a direct collision as shown in Figure 4.6.

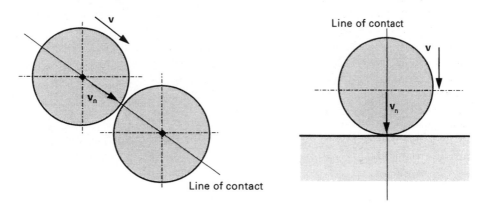

Figure 4.6   Direct collisions

But the analysis is made more complicated when the impact is *oblique*, as shown in Figure 4.7. The normal velocity for an oblique collision is the component of velocity acting along the line of contact. For two spheres, the line of contact joins their centres; for anything colliding with a flat surface, the line of contact is always perpendicular to that surface.

For oblique collisions there are always two components of velocity: v_n the normal velocity, and v_t the tangential velocity.

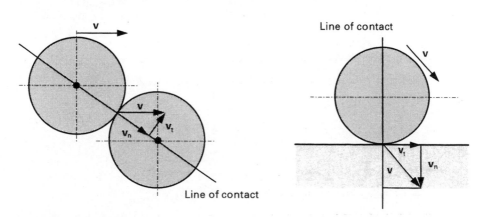

Figure 4.7   Oblique collisions

Example 4.5   *A ball is dropped onto a hard concrete floor. The speed of the ball is measured just prior to and just after hitting the floor and found to be 4 m/s and 2 m/s respectively. Determine the coefficient of restitution for the ball colliding with concrete.*

SOLUTION
First we assume that the ground has zero velocity before and after impact (later we shall see that this is in fact wrong but the error incurred is very small). We also sketch the problem as shown in Figure E4.5.

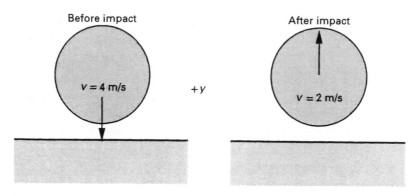

Figure E4.5

If we name the ball A and the ground B we can use equation (4.4), and by rearranging $(v_b - v_a)' = e(v_a - v_b)$ we get

$$e = \frac{(v'_b - v'_a)}{(v_a - v_b)} = \frac{(0 - 2)}{(-4 - 0)} = 0.5$$

4.3   Conservation of momentum

The principle of conservation of momentum is one of the fundamental laws of physics. As a tool, engineers find it invaluable. The principle of conservation of momentum states:

The total momentum within a system remains constant.

The principle may be exemplified by considering two spherical bodies (A and B) colliding as shown in Figure 4.8.

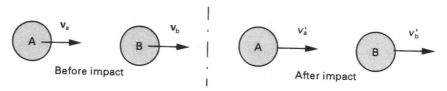

Figure 4.8   Two bodies colliding

The total momentum in the system before impact is given by

$$\mathbf{p} = (m_a\mathbf{v}_a) + (m_b\mathbf{v}_b)$$

If the mass of each object remains the same, after the impact the total momentum is given by

$$\mathbf{p}' = (m_a\mathbf{v}'_a) + (m_b\mathbf{v}'_b)$$

The principle states that total momentum must remain constant hence $\mathbf{p} = \mathbf{p}'$, or

$$(m_a\mathbf{v}_a) + (m_b\mathbf{v}_b) = (m_a\mathbf{v}'_a) + (m_b\mathbf{v}'_b) \tag{4.5}$$

Note that in equation (4.5) the individual momentums of objects A and B change, but the overall value of momentum does not. Hence a *system* consists of several items, not just one, and not always only two.

Example 4.6 *Two billiard balls collide, as demonstrated in Figure E4.6. Before the collision, ball A is travelling at 1.5 m/s and ball B is stationary. After the collision, ball A travels in the same direction but at 0.6 m/s. If ball A has a mass of 0.3 kg and ball B has a mass of 0.35 kg, determine the speed of ball B after impact. Also determine the impulse which acted on ball B.*

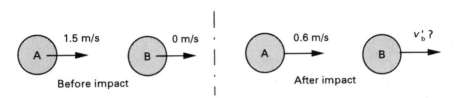

Figure E4.6

SOLUTION
All the balls are moving in the same direction, so we can use equation (4.5) in scalar form:

$$(m_a v_a) + (m_b v_b) = (m_a v'_a) + (m_b v'_b)$$

Identifying known quantities

$$m_a = 0.3\,\text{kg}$$
$$m_b = 0.35\,\text{kg}$$
$$v_a = 1.5\,\text{m/s}$$
$$v_b = 0$$
$$v'_a = 0.6\,\text{m/s}$$

we therefore need to determine v'_b, hence

$$(0.3 \times 1.5) + (0.35 \times 0) = (0.3 \times 0.6) + (0.35 \times v'_b)$$
$$0.45 = 0.18 + 0.35v'_b$$

or

$$v'_b = (0.45 - 0.18)/0.35 = 0.77\,\text{m/s}$$

Impulse is defined as the change of momentum of a body. Before impact the momentum of B is zero. After impact it is (0.35×0.77), hence

$$\Delta p = (0.35 \times 0.77) - 0 = 0.27 \, \text{kg m/s}$$

Example 4.7 *Why do skaters spin faster when they draw their arms in from being extended to a tight tuck?*

SOLUTION

The reason for this increase in speed is because of conservation of momentum. With your arms extended your moment of inertia is greater than when they are drawn in. Remembering that angular momentum is given by $I\omega$, we have

$$I_1 \omega_1 = I_2 \omega_2$$

If I_2 is less than I_1, ω_2 must be greater than ω_1. Figure E4.7 attempts to demonstrate this phenomenon.

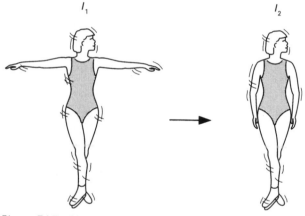

Figure E4.7   Skater

You may wish to try out an example of the conservation of momentum that can be readily attempted. All you have to do is find a swivel chair. Sit on the chair with your legs extended and get another person to spin you around freely; you will spin with an angular velocity ω_1. If you draw your legs in towards your body, you will notice that your angular velocity increases to ω_2.

Example 4.8 *Two identical spheres impact as illustrated in Figure E4.8. Sphere A is moving horizontally with an initial velocity of $v_a = 4 \, m/s$, and sphere B is stationary. Given that the coefficient of restitution between the two spheres is 0.6, determine their respective velocities after impact. You may assume frictionless contact.*

SOLUTION

First we need to determine the normal velocity of impact. We need to add some axes to Figure E4.8(a), orienting the x-axis with the line of contact makes the solution easier.

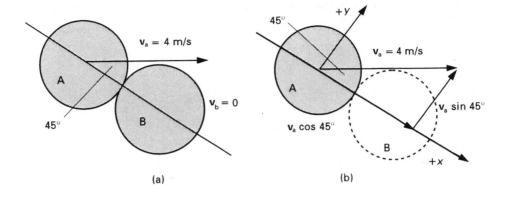

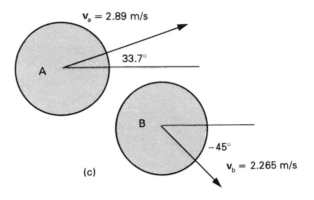

Figure E4.8

Using equation (4.4) in the x-direction reveals that

$$(\mathbf{v}'_{bx} - \mathbf{v}'_{ax}) = e(\mathbf{v}_{ax} - \mathbf{v}_{bx})$$

the added subscript x reminds us that we are working in the x-direction only. When we substitute in our known values this becomes

$$(\mathbf{v}'_{bx} - \mathbf{v}'_{ax}) = 0.6 \times (4\cos 45^\circ - 0)$$
$$= 0.6 \times 2.83 \qquad\qquad (1)$$
$$(\mathbf{v}'_{bx} - \mathbf{v}'_{ax}) = 1.7\,\text{m/s}$$

which states that after impact the difference in velocity *along* the x-axis between spheres B and A is 1.7 m/s. But we do not have actual velocities. To determine them, we recall the principle of conservation of momentum (in the x-direction only):

$$(m_a \mathbf{v}_{ax}) + (m_b \mathbf{v}_{bx}) = (m_a \mathbf{v}'_{ax}) + (m_b \mathbf{v}'_{bx})$$

which can be reduced because $m_a = m_b$

$$m(\mathbf{v}_{ax} + \mathbf{v}_{bx}) = m(\mathbf{v}'_{ax} + \mathbf{v}'_{bx})$$

Dividing through by m gives

$$\mathbf{v}_{ax} + \mathbf{v}_{bx} = \mathbf{v}'_{ax} + \mathbf{v}'_{bx}$$

Now we substitute in our known values

$$2.83 + 0 = v'_{ax} + v'_{bx}$$

We also know, from (1), that

$$(v'_{bx} - v'_{ax}) = 1.7 \, \text{m/s}$$

or

$$v'_{bx} = 1.7 + v'_{ax}$$

hence

$$2.83 + 0 = v'_{ax} + 1.7 + v'_{ax}$$
$$2.83 = 2v'_{ax} + 1.7$$
$$v'_{ax} = (2.83 - 1.7)/2 = 0.565 \, \text{m/s}$$

thus

$$v'_{bx} = 1.7 + 0.565 = 2.265 \, \text{m/s}$$

If we assume that the balls are frictionless, the tangential velocities of A and B remain unchanged, thus

$$v'_{ay} = 4\sin 45° = 2.83 \, \text{m/s}$$
$$v'_{by} = 0$$

This means that momentum in the *y*-direction is conserved!
Hence

$$v_a = 0.565i + 2.83j = 2.89 \text{ at } 78.7°$$
$$v_b = 2.265i = 2.265 \text{ at } 0°$$

with reference to the *x*-axis as drawn in Figure E4.8(b). If we want to convert them to refer to our horizontal, we remember that our *x*-axis is −45° to the horizontal and we add −45° to all the angles above:

$$v_a = 2.89 \text{ at } 33.7°$$
$$v_b = 2.265 \text{ at } -45°$$

4.4   Newton's laws of motion

A major proportion of engineering mechanics is based on the three laws proposed by Newton. These laws are so fundamental that their basic principles, not the actual wording, should be fully understood. From this text's viewpoint they also define *force*.

First Law

A body remains in the same state of rest or uniform linear motion unless acted upon by an external force.

Second Law

The rate of change of momenum of a body is proportional to the applied force and is in the direction of the applied force.

Third Law
Every action has an equal and opposite reaction.

To an engineer it is the second law which is used as a basis for engineering mechanics. The other two have a role but are more often used intuitively rather than specifically. Let us consider each law in turn.

First Law
Consider a football lying static on a football pitch. The goalkeeper kicks the ball and it flies in the direction of the centre of the pitch. In other words, the foot of the goalkeeper applied a force to the ball which changed its state from being at rest to being in motion. The first law simply states that for something to happen, something must make it happen – and that something is a *force*.

Second Law
The second law expands on the first by considering what happens to a body due to an action. At first the ball (described above) is stationary, then it is in motion. The foot has applied an impulse to the ball and the ball has accelerated; there has been a change of momentum. Furthermore, the direction of travel of the ball is in the direction of the kick. Thus the second law provides a link between force and motion.

Third Law
The third law can be exemplified by what the goalkeeper feels when the ball is kicked. In other words the ball has 'applied' as much force to the foot as the foot did to the ball – equal and opposite. If the ball had been made from concrete ... well, you don't need a textbook to explain that one!

A close examination of the second law defines force. Rate of change of momentum is proportional to force, therefore

$$\mathbf{F} \propto \frac{\mathbf{p}_2 - \mathbf{p}_1}{t_2 - t_1} \quad \text{or} \quad \frac{d\mathbf{p}}{dt} \tag{4.6}$$

If we consider mass as a constant this becomes

$$\mathbf{F} \propto \frac{m\mathbf{v}_2 - m\mathbf{v}_1}{t}$$

or

$$\mathbf{F} \propto m\left(\frac{\mathbf{v}_2 - \mathbf{v}_1}{t}\right)$$

which we recognise as

$$\mathbf{F} \propto m\mathbf{a}$$

We need to eliminate the proportionality. To do this we assign the newton (N) to be the unit of force. We further define 1 newton to be the force required to accelerate 1 kg by 1 m/s². Hence

$$\mathbf{F} = m\mathbf{a} \tag{4.7}$$

Note that force is a vector quantity. Of course, the mass of an object does not have to remain constant (e.g. in a rocket). Then equation (4.6) becomes

$$F = \frac{d(m\mathbf{v})}{dt}$$

Systems where mass is not constant are not as easy to model as those with constant mass.

Example 4.9 *A car of mass 1.23 tonnes accelerates from 35 km/h to 75 km/h in 25 s. Determine the force required to have achieved this acceleration.*

SOLUTION
Converting to SI units:-

$$v_1 = 35/3.6 = 9.72 \text{ m/s}$$
$$v_2 = 75/3.6 = 20.83 \text{ m/s}$$

Assuming constant acceleration, we can use equation (3.1) to get

$$a = (v_2 - v_1)/t = (20.83 - 9.72)/25 = 0.44 \text{ m/s}^2$$

Recalling equation (4.7)

$$F = ma = 1230 \times 0.44 = 546 \text{ N}$$

4.5   Application of the second law to engineering

The second law as defined by equation (4.6) is used by engineers in two forms. Both forms come from an examination of the reduced equation (4.7), $F = ma$.

If a body is not moving, or is not expected to move, the acceleration of that body is zero. It is said to be *static*. Hence equation (4.7) can be reduced to

$$F = 0$$

However, there is commonly more than one force acting on a body, therefore it is the resultant force which applies (remember force is a vector):

If a body is static the sum of the forces is zero.

Mathematically

$$\Sigma F = 0 \tag{4.8}$$

where the symbol Σ means 'sum of'. The study of bodies, objects or systems where equation (4.8) applies is called *statics*.

If a body is in motion (in particular, motion which involves acceleration) then equation (4.7) stands:

The motion of a body is governed by the resultant force acting upon it.

Mathematically

$$\Sigma F = ma \tag{4.9}$$

The study of bodies, objects or systems to which equation (4.9) applies is called *dynamics*.

Associated with equation (4.9) we can use equation (4.6) directly by considering impulse. Impulse is defined as the change of momentum ($\Delta \mathbf{p} = \mathbf{p}_2 - \mathbf{p}_1$), so we can also determine the force acting on a body by considering momentum:

$$\mathbf{F} = \Delta \mathbf{p}/t \qquad (4.10)$$

or, impulse = force × time; this explains why another accepted unit for momentum and impulse is the newton-second (N s).

Example 4.10 *In a soccer match a goalkeeper kicks a ball from rest. The ball has a mass of 1.5 kg and a velocity of 21 m/s after it has been kicked. If the foot is in contact with the ball for a period of 0.3 s, determine the average force exerted over this period.*

SOLUTION
We recognise this as a force–acceleration problem. We can also approach it using impulse. The change in momentum of the ball (assuming no change of direction throughout) is

$$\Delta p = 1.5 \times (21 - 0)$$
$$= 31.5\,\text{N s}$$

Hence from equation (4.10)

$$F = \frac{31.5}{0.3}$$
$$- 105\,\text{N}$$

This problem highlights the importance of follow-through. If the applied force is constant throughout then the longer the foot is in contact with the ball, the greater will be the change in momentum!

4.5.1 Centrifugal force

In Chapter 2 we met centripetal acceleration, the acceleration required to force an object to travel in a circular path, instead of a straight path. We now know that $F = m\mathbf{a}$, therefore the centripetal acceleration must be caused by a force, as illustrated in Figure 4.9. Since the centripetal acceleration is given by

$$a_c = \omega^2 r$$

From Newton's second law, the centripetal force exerted on the body is given by

$$F_c = ma_c$$
$$= mr\omega^2 \qquad (4.11)$$

or

$$F_c = m\frac{v^2}{r}$$

However, Newton's third law states that there is also an equal and opposite force, and this is called *centrifugal force*. The centrifugal force (F_c) acts in the opposite direction to the centripetal force, hence radially outwards. If the body is hollow, any loose objects within the body will be subject to this force and will be 'thrown'

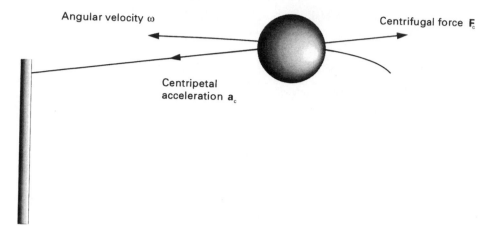

Angular velocity ω

Centrifugal force F_c

Centripetal
acceleration a_c

Figure 4.9   Centripetal acceleration and centrifugal force

outwards, as will any further objects attached to the outer surface of the body. It is the centrifugal force we commonly associate with rotation, as exhibited in blood centrifuges and funfair rides.

Example 4.11   *A car of mass 1200 kg negotiates a bend of radius 400 m with a constant speed of 72 km/h. Determine the centrifugal force acting on the car, and state its effect.*

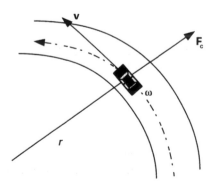

v

F_c

ω

r

Figure E4.11

SOLUTION
A quick sketch of a car going around a bend (Figure E4.11), viewed from above, illustrates that it is no different from any other object moving in an arc. The angular velocity of the car is determined from

$$v = \omega r$$

hence

$$\omega = \frac{v}{r}$$

$$= \frac{(72/3.6)}{400}$$

$$= 0.05 \, \text{rad/s}$$

From equation (4.11)

$$F_c = mr\omega^2$$

$$= 1200 \times 400 \times 0.05^2$$

$$= 1200 \, \text{N}$$

This force is directed radially outwards, so it is tending to force the car off the road. It also has the annoying effect of throwing cassette boxes off the dashboard!

4.6 Forces and moments

We have already seen that objects prefer to move in straight lines. We have now learnt that to make an object move in a straight line requires a force. So what makes an object, such as a tap or a nut, spin or twist? Here we have separate terminology for actions producing linear motion and actions inducing angular motion.

Consider the action of tightening a wheel-nut on an automobile using a wrench (as illustrated in Figure 4.10)

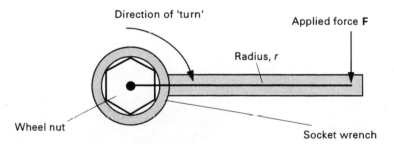

Figure 4.10   Moment applied due to a force acting over a distance

Because the force applied acts about a centre of rotation, and also because it is at a radius r from the centre of rotation, the force causes the nut to turn (in this case clockwise). The product of force and radius, or *moment arm*, is called *moment*:

$$M = Fr \qquad\qquad (4.12)$$

It is measured (vector form $\mathbf{M} = \mathbf{F}r$) in newton-metres (N m) and the sign convention is anticlockwise = positive because an anticlockwise turn causes a positive rotation. The sense of rotation can be evaluated by inspection or by vector algebra, but vector algebra is beyond the scope of this text. Another name for moment is *torque* (T), but this is often reserved for rotating machines such as internal combustion engines, electric motors and driveshafts. Torque is also used as a measure for 'tightness' of a nut or bolt.

Newton's second law may be written with respect to angular motion as well as linear motion.

For a static system, the sum of the moments must equal zero, hence

$$\Sigma M = 0 \tag{4.13}$$

For a dynamic system, the sum of the moments is equal to the product of moment of inertia and angular acceleration hence

$$\Sigma M = I\alpha \tag{4.14}$$

Summary

Momentum
Linear: $\mathbf{p} = m\mathbf{v}$ (kg m/s or N s)
Angular: $\mathbf{L} = I\omega$ (kg m^2/s or N m s)

Impulse
$$\Delta \mathbf{p} = \mathbf{p}_2 - \mathbf{p}_1 = m_2\mathbf{v}_2 - m_1\mathbf{v}_1$$
$$\mathbf{F} = \Delta \mathbf{p}/t$$

Impacts
$$(\mathbf{v}_b' - \mathbf{v}_a') = e(\mathbf{v}_a - \mathbf{v}_b)$$

where e = coefficient of restitution

$e = 0$ $\Rightarrow$ plastic

$e = 1$ $\Rightarrow$ elastic

$0 < e < 1$ $\Rightarrow$ real

Conservation of momentum
$$(m_a\mathbf{v}_a) + (m_b\mathbf{v}_b) = (m_a\mathbf{v}_a') + (m_b\mathbf{v}_b')$$

Newton's laws of motion
First law: A body remains in the same state of rest or uniform linear motion unless acted upon by an external force.
Second law: The rate of change of momentum of a body is proportional to the applied force and is in the direction of the applied force.
Third law: Every action has an equal and opposite reaction.

Force
$$\mathbf{F} = ma \quad (\text{N})$$

Centrifugal force

$$F_c = mr\omega^2 \quad \text{(N) (direction radially outwards)}$$

Statics

$$\Sigma F = 0$$
$$\Sigma M = 0$$

Dynamics

$$\Sigma F = ma$$
$$\Sigma M = I\alpha$$

Applied moment

$$M = Fr \quad \text{(N m)}$$

Problems

Momentum and impulse

4.1   A car of mass 1.25 tonnes travels at a constant speed of 120 km/h. Determine the momentum of the car.

4.2   A train of mass 300 tonnes climbs a hill of gradient 20° at a speed of 75 km/h. Determine

    (a)   the momentum of the train
    (b)   the horizontal component of momentum
    (c)   the vertical component of momentum

4.3   A wheel of moment of inertia 12 kg m² rotates clockwise at 3600 rev/min. Determine the angular momentum of the wheel, also determine the sense of the vector.

4.4   A trolley of mass 150 kg runs on wheels of 250 mm diameter and moment of inertia 1.5 kg m². If the speed of the trolley is 25 km/h, determine the linear momentum of the trolley and the angular momentum of each wheel.

4.5   A flywheel of diameter 350 mm rotates at a constant speed of 450 rev/min. If the moment of inertia of the flywheel is 3.4 kg m², determine

    (a)   the peripheral velocity of the flywheel
    (b)   the angular momentum of the flywheel

Conservation of momentum and impacts

4.6   The velocity of a 2 kg object was 25 m/s before a collision with a second object. After the collision the velocity was 15 m/s. Determine the impulse on the first body, hence determine the impulse on the second object.

4.7   A ball hits a heavy, hard surface with a normal velocity of 17 m/s. If the ball bounces off the surface with a velocity of 12 m/s, but in the opposite direction, determine the coefficient of restitution for the collision.

4.8   A box of mass 3 kg is thrown with a velocity of 10 m/s into a trolley of mass 17 kg. The box is thrown horizontally and remains within the trolley. Determine the velocity of the bodies after the impact.

4.9   An object of mass 2 kg impacts with a stationary object of mass 4 kg. The collision is normal. The velocity of the 2 kg mass is 2 m/s before impact and the coefficient of restitution is 0.5. Determine the velocities of both objects after the impact.

4.10   An object of mass 3 kg has an intial velocity of 3 m/s prior to an impact with a second stationary object. After the impact, the first object's velocity is 0.5 m/s and the second object's is 2.5 m/s. Determine
(a)   the coefficient of restitution for the impact
(b)   the inital momentum of both objects
(c)   the impulse acted on the first object
(d)   the mass of the second object

4.11   A ball is dropped from a height of 2 m on to hard ground. The coefficient of restitution for the impact is 0.6. Determine
(a)   the velocity of the ball just prior to impact
(b)   the velocity of the ball just after impact
(c)   the height the ball reaches after the first bounce

4.12   Determine the height reached by the ball in Problem 4.11; do this for each bounce up to seven impacts. What is wrong with the model?

4.13   A 2 kg object travels with a velocity of 2 m/s horizontally. It impacts with a stationary 2 kg object; the line of contact is at an angle of $-30°$ to the horizontal. If after the impact the velocity of the second object is 1.5 m/s at $-30°$, determine the velocity of the first object and the coefficient of restitution for the impact.

Forces and moments

4.14   Define the term *impulse*. How does this relate to force as defined by Newton's second law of motion?

4.15   State Newton's three laws of motion. How is the second law interpreted for statics, and for dynamics.

4.16   A car of mass 1.25 tonnes travels at a speed of 120 km/h. The car's brakes are applied for a period of 10 s, after which time the car's speed is reduced to 60 km/h. Determine
(a)   the change in momentum of the car
(b)   the braking force applied (assuming it is constant)

4.17   A force of 2.5 kN is applied to one end of an arm of length 1.25 m, and the other end is attached to a wall. Determine the moment applied by the force at the fixed end.

4.18   A fairground carousel of diameter 10 m rotates at a constant speed of 25 rev/min. Determine
(a)   the peripheral speed of the carousel
(b)   the centripetal acceleration acting on a body placed at the periphery
(c)   the magnitude of the centrifugal force acting on a 75 kg person standing 3.5 m from the centre

4.19   A block of mass 2 kg is placed on the end of an arm 2 m long. The arm is then spun about one end at a speed of 300 rev/min. Determine
(a)   the centripetal acceleration acting on the block
(b)   the magnitude and direction of the force acting on the block

4.20   A flywheel of moment of inertia 17 kg m^2 rotates at a constant speed of 2000 rev/min. A constant braking torque is applied for a period of 10 s, after which time the speed of the flywheel is 50 rev/min. Determine
(a)   the acceleration of the flywheel
(b)   the braking torque applied
(c)   the number of rotations of the flywheel during this period

4.21 A car of mass 800 kg travels at a constant speed of 75 km/h. A braking force is applied, represented by the force-time data given in Figure P4.20. Determine the speed of the car (in km/h) after the braking period. Also state its change in momentum.

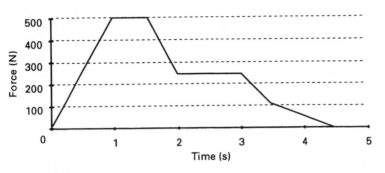

Figure P4.20

Chapter 5

Work, energy and power

We have already met some of the fundamental units which are used by engineers to model real-life problems. The last set of units we need to examine are the units of work, energy and power. In this chapter we shall also meet another fundamental principle to add to our toolbox, the principle of conservation of energy.

5.1 Work and energy

In Chapter 4 we saw that we could not change the state of an object without applying an impulse. We also saw that the application of a force is required to change a body's momentum. What we did not see was that all of this 'change' has to be undertaken, that is something has to 'do it'.

For a change in state to have occured, it is said that *work* has been done. Work is defined as

the product of applied force and displacement of the force, where the displacement is in the same direction as the applied force.

We recall (Section 1.4.4) that when two vectors are aligned the result of the product is a scalar quantity. Displacement and force are both vectors, therefore work is a scalar. For a constant force F we can define work by

$$W = Fs \tag{5.1}$$

For any arbitrary force the general equation for work is

$$W = \int_0^s F ds$$

which we recognise as the area under the force–displacement graph (as illustrated in Figure 5.1).

The symbol for work is W. By inspection the unit would seem to be N m, however, work has a specific unit, the joule (J).[1] One joule (1 J) is defined as the work required to move a force of 1 N by 1 m.

---

[1] After British physicist James Prescott Joule (1818–1889).

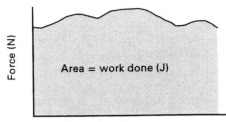

Figure 5.1   Work is the area under the force–displacement graph

We have now seen that to change a system we have to do work. But the question arises, Where does the work come from? To answer this question fully would require journeying back in time to the Big Bang. We need only realise that stored within a body or a system is the potential to do work, this is *energy*. Hence work and energy are, effectively, the same.

Thus work can be thought as *what has been done* and energy can be thought of as *what can be done*. For this reason work can have a sign, since something can either do work (negative) or have work done to it (positive). The amount of energy in a system is absolute. It can only be positive. The change in energy of a system can, of course, have any sign.

Example 5.1   *A box of mass 250 kg lies on a concrete floor. Two people push the box a distance of 2 m, and the force required to move the box is constant at 1250 N. Determine the work done to move the box.*

SOLUTION
Recalling equation (5.1)

$$W = Fs = 1250 \times 2$$
$$= 2500 \, \text{J} \quad \text{or} \quad 2.5 \, \text{kJ}$$

The work that is done on, or to, a body can be from several different energy sources. Let us examine in more detail the various forms which energy can take.

5.1.1   Gravitational potential energy
Gravitational potential energy is the energy stored within a body due to its position in a gravitational field. In everyday life the gravitational field is that of the earth and the position is height. However one must remember that the engineer does not only work on terra firma! Gravitational potential energy is often shortened to potential energy and is assigned the symbol E_p.

The gravitational potential energy of a body is defined as the product of mass, gravitational acceleration and height from an arbitrary datum:

$$E_p = mgh$$

where m is the mass of the body, g is gravitational acceleration ($9.81\,\text{m/s}^2$) and h is the height of the object above a known datum.[2]

If the value of h is very large, the value of gravitational acceleration changes. In this case the general expression for gravitational potential energy is drawn from Newton's law of gravitation (Appendix D).

The change in potential energy (or work done against gravity) when an object moves from one position (mgh_1) to another (mgh_2) is given by

$$\Delta E_p = (mgh)_2 - (mgh)_1 \tag{5.3}$$

This is illustrated in Figure 5.2.

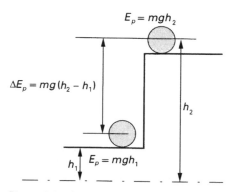

$E_p = mgh_2$

$\Delta E_p = mg(h_2 - h_1)$

h_2

h_1 $E_p = mgh_1$

Figure 5.2   Graphical representation of potential energy

If the result of (5.3) is positive, work is said to have been done on the object. If the result is negative, potential energy has been lost from the object. But if the difference between h_2 and h_1 is very large (see Example 5.3), gravity cannot be considered to be constant.

Example 5.2   *Determine the potential energy stored in a box of mass 52 kg when it has been pushed up a slope of length 30 m and angle 25°. Determine the work done against gravity to carry out this task.*

SOLUTION

It is important to remember that potential energy depends on height, as illustrated in Figure E5.2. It is immaterial how that height is achieved. Hence at the top of the slope the potential energy of the box (relative to the base of the slope) is given by

$$E_p = mgh = 52 \times 9.81 \times 30 \sin (25°) = 6.47\,\text{kJ}$$

The work done against gravity is also 6.47 kJ.

---

[2] Some texts use the symbol V for potential energy.

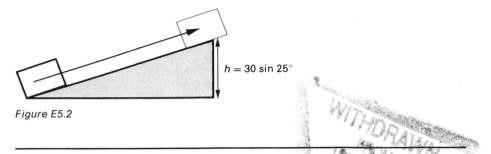

Figure E5.2

Example 5.3 *A satellite of mass 125 kg orbits the earth at a height of 50 km (Figure E5.3). Determine the potential energy of the satellite with respect to the surface of the earth.*

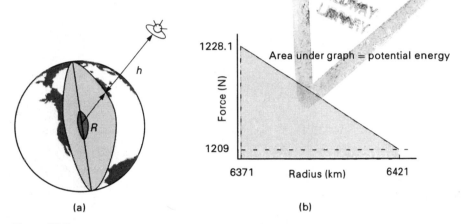

(a)

(b)

Figure E5.3

SOLUTION

First we must recognise that the distances here are very great, so we cannot use $g = 9.81 \text{ m/s}^2$.

Let us determine g using Newton's law of gravitation (see Appendix D):

$$g = G\frac{m_e}{(R+h)^2} = 6.673 \times 10^{-11} \frac{5.976 \times 10^{24}}{(6371 \times 10^3 + 50 \times 10^3)^2} = 9.67 \text{ m/s}^2$$

Comparing this to our normal gravitational acceleration, $g = 9.81 \text{ m/s}^2$, shows a slight difference. So we should draw a graph of mg versus h and determine the area underneath (as illustrated in Figure E5.3(b)). The area can thus be estimated:

E_p = area under mg v h graph

= 60.93 MJ

But since the change in g is so small over this distance, we could estimate potential energy using an average value of $g = 9.74 \text{ m/s}^2$:

$E_p = mgh$

= $125 \times 9.74 \times 50 \times 10^3$

= 60.88 MJ

5.1.2   Kinetic energy

5.1.2.1   *Linear motion*

Consider a body of mass m moving at constant speed v. If a constant force F, resisting motion, is applied to the body (as illustrated in Figure 5.3), it will eventually stop the body.

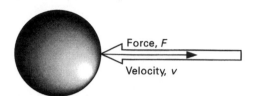

Figure 5.3   A body subjected to an opposing force F

Recalling equation (4.7) in scalar form

$$F = ma$$

or

$$a = \frac{F}{m}$$

which will be constant. We also know that, before the force is applied, $v_1 = v$. After some time interval the object stops, so $v_2 = 0$. For a system acting under constant acceleration (Chapter 3), we have

$$v_2^2 = v_1^2 + 2as$$

or

$$0 = v^2 + 2\left(\frac{F}{m}\right)s$$

which we can rearrange for s, the distance the object moves through before it comes to rest. This gives

$$s = -\frac{1}{2}\left(\frac{m}{F}\right)v^2$$

We now recall equation (5.1), which shows that work done is the product of force and distance, hence

$$W = Fs$$

which, when we substitute in our value of s, becomes

$$W = F \times \left(-\frac{1}{2}\frac{m}{F}\right)v^2 = -\frac{1}{2}mv^2$$

This is the work done to stop the mass by the force F, which is the same as the energy stored within the mass initially. This energy is called *kinetic energy*; it is the energy stored in a body due to motion. Kinetic energy is given the symbol E_k[3] where

$$E_k = \tfrac{1}{2}mv^2 \tag{5.4}$$

Example 5.4 *An automobile of mass 1250 kg travels at a constant speed of 56 km/h. Determine the kinetic energy of the vehicle.*

SOLUTION
Using equation (5.4), we have

$$E_k = \tfrac{1}{2}mv^2$$
$$= \tfrac{1}{2} \times 1250 \times (56/3.6)^2$$
$$= 151.235\,\text{kJ}$$

Angular motion
Consider a body of moment of inertia I, rotating about an axis with angular velocity ω (Figure 5.4). If a resisting torque, T, is applied to resist motion, the body will eventually come to rest.

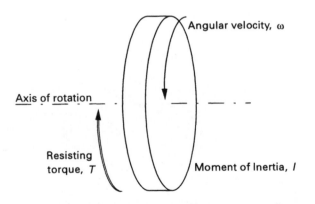

Figure 5.4   Kinetic energy of a rotating body

Using a similar proof as when we obtained equation (5.4), we can derive an expression for kinetic energy stored within a rotating body. This is given by

$$E_k = \tfrac{1}{2}I\omega^2 \tag{5.5}$$

---

[3] Some texts use the symbol T for kinetic energy; this may be confused with temperature and torque, so this text uses E_k.

Example 5.5   *A round barrel of mass 45 kg and moment of inertia 12 kg m² rolls along a flat surface with a speed of 2 m/s. If the barrel's diameter is 400 mm, determine*

    (a) *the linear kinetic energy of the barrel*
    (b) *the angular kinetic energy of the barrel*
    (c) *the total kinetic energy of the barrel*

SOLUTION

All three parts of this problem stem from an earlier example, Example 2.6. If the barrel rolls without slipping, the linear speed of the barrel is the same as the peripheral speed of the barrel, and if it is not slipping it must be rotating (see Figure E5.5). In this instance the object has both a linear and an angular velocity. Thus the barrel has both linear *and* angular kinetic energies.

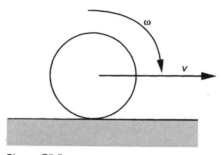

Figure E5.5

Thus

    (a) Linear E_k

$$E_k = \tfrac{1}{2}mv^2 = \tfrac{1}{2} \times 45 \times 2^2 = 90\,\text{J}$$

    (b) Angular E_k

$$E_k = \tfrac{1}{2}I\omega^2$$

but since $\omega = v/r$, we get

$$E_k = \tfrac{1}{2}I(v/r)^2 = \tfrac{1}{2} \times 12 \times (2/0.2)^2 = 600\,\text{J}$$

    (c) The total kinetic energy of the barrel is the sum of linear and angular kinetic energy:

$$E_{k,\text{total}} = E_{k,\text{linear}} + E_{k,\text{angular}} = 90 + 600 = 690\,\text{J}$$

5.1.3   *Stored energy in springs*

Springs are energy storage devices used in engineering. They have been used to drive clocks, watches and toys, for example.

When a spring is compressed or extended, (Figure 5.5) its *stiffness* is defined by *the gradient of the line generated by plotting applied force versus displacement* (Figure 5.6).

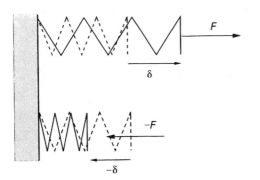

Figure 5.5   Extension and compression of a spring due to a force *F*

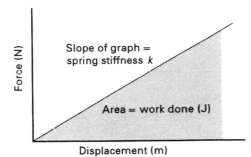

Figure 5.6   Force–displacement relationship for a spring

Thus spring stiffness k is defined by

$$k = \frac{F}{\delta} \tag{5.6}$$

where the displacement δ of the spring is in the direction of the applied force F. Recalling from Section 5.1 that the work done is the area under the force–displacement graph, it is evident that the work done on the spring by extension or compression is

$$W = \tfrac{1}{2}F\delta$$

Notice that a factor of 1/2 has been introduced because the force is not constant. The increase in force is directly proportional to the displacement of the spring. If we rearrange equation (5.6) we get

$$F = k\delta$$

and hence

$$W = \tfrac{1}{2}k\delta^2 \tag{5.7}$$

The work done on a spring is stored in the form of *stored elastic* energy. The general symbol for stored elastic energy is E_e and its value is given by equation (5.7).

Example 5.6   *A spring of stiffness 500 N/m is compressed from its free length by 50 mm. Determine*

(a) *the force applied to the spring*
(b) *the energy stored within the spring*

SOLUTION
The *free length* of a spring is a name attributed to the spring in its natural state. This length changes when a load is applied.

(a)   Rearranging equation (5.6) we obtain

$$F = k\delta$$

hence

$$F = 500 \times 0.05$$
$$= 25 \, \text{N}$$

(b)   Using equation (5.7), the work done on the spring is given by

$$W = \tfrac{1}{2}k\delta^2$$

which is the same as the energy stored within the spring (U), hence

$$E_e = \tfrac{1}{2} \times 500 \times 0.05^2$$

thus

$$E_e = 0.625 \, \text{J}$$

Example 5.7   *Figure E5.7 illustrates a support for an electric motor which comprises four springs of stiffness 0.5 kN/m. If the motor has a mass of 20 kg, determine*

(a) *the static deflection of the springs*
(b) *the total stored elastic energy.*

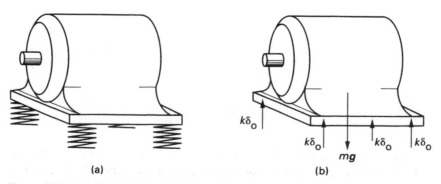

(a)   (b)

Figure E5.7   (a) Actual motor and (b) its free-body diagram

SOLUTION

(a) A free-body diagram (Figure E5.7(b)) of the electric motor illustrates that each spring must carry 1/4 of the body force, if we assume that the centre of gravity is exactly central.

Applying Newton's second law (for statics)

$$\sum F = 0$$
$$4k\delta_o - mg = 0$$

or, on rearranging for δ_o, the static deflection of the spring (static deflection is the deflection of a spring, or any supporting structure, due to a body force), is

$$\delta_o = mg/4k$$
$$= (20 \times 9.81)/(4 \times 500)$$
$$= 0.0981\,\text{m} = 98.1\,\text{mm}$$

The position in space adopted by a body supported by another structure is often called its *static equilibrium* position.

(b) The total stored energy will be the summation of the energy stored in each spring, so

$$E_e = 4 \times \left(\tfrac{1}{2}k\delta^2\right)$$
$$= 4 \times \left(\tfrac{1}{2} \times 500 \times 0.0981^2\right)$$

hence

$$E_e = 9.62\,\text{J}$$

5.1.4 *Heat*

The forms of energy transfer we have just described occur as a result of mechanical intervention – a force or torque being applied to a system. Heat is different because there is no 'mechanical' intervention. The energy may be transferred from one body to another by three mechanisms: *conduction*, *convection* and *radiation* (Figure 5.7). Energy transfer by heat is given the symbol Q (as opposed to W for work) and has the same unit, the joule (J).

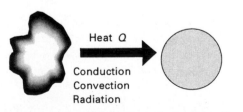

Heat Q

Conduction
Convection
Radiation

Figure 5.7 Transfer of energy in the form of heat

James Prescott Joule showed that there is an equivalence between mechanical energy (work) and heat. In other words, the temperature of a cup of water can be raised by transfer of energy in the form of heat or by stirring. This was the basis of his famous experiments. The amount of energy required to change the temperature of a body is given by

$$Q = mc\Delta T \tag{5.8}$$

where

m = the mass of the body

c = the specific heat capacity of the body's material

ΔT = the temperature change $(T_2 - T_1)$

Remember that equation (5.8) can be as a result of work or heat (W or Q).

The specific heat capacity of a body is a material constant. A material with a high specific heat capacity requires more heat to raise its temperature. A material with a low specific heat capacity require less heat. Strictly speaking, the specific heat capacity should be defined with respect to *constant pressure* c_p, or *constant volume* c_v. We will assume a constant-pressure system, since this corresponds to the heating of solids and liquids in open atmosphere. In Chapter 8 we shall meet equation (5.8) again, but there we shall generalise it to cope with any fluid.

Example 5.8 *During a manufacturing process a 15 kg steel bar must be heated, prior to final machining, to a temperature of 840 °C. If the nominal temperature before heating was 23 °C, determine the heat required to achieve this temperature change. Assume that* $c_p = 446 \, J/kg\,K$ *for this grade of steel.*

SOLUTION
First we note equation (5.8), where heat Q is defined as

$$Q = mc\Delta T$$
$$Q = 15 \times 446 \times (840 - 23)$$

hence

$$Q = 5.47 \, MJ$$

5.1.5 Energy in real terms.
We have met the joule (J), which we have defined as the work done to move 1 N by 1 m. But what does this mean on a scale of real-life events? Table 5.1 illustrates some occurrences with their associated energies. Some of the events you meet every day, but some I hope you never meet. As you will notice, the range is staggering.

5.2 Conservation of energy

The *principle of conservation of energy* is a fundamental law of physics:

Energy cannot be created or destroyed. It can only be transferred from one form to another. The total energy within a system remains constant.

Table 5.1 Some typical energy ratings for common events

Event	Energy
The beating of a fly's wing	$10\,\mu J$
The energy of visible light emitted by a domestic light-bulb in 1 s	10 J
The kinetic energy of a rifle bullet	100 J
The potential energy gained by climbing the stairs	1.5 kJ
The kinetic energy of a family saloon on a motorway	734 kJ
The energy used to heat a bath of water	10 MJ
The kinetic energy of Concorde at Mach 2	10 GJ
The energy released by an earthquake	100×10^{18} J
The energy released by an exploding star (supernova)	1×10^{44} J

Consider the solution to Example 4.6, illustrated in Figure 5.8. Let us now check the energies before and after impact. The bodies are moving with linear velocity, so the form of energy to consider is kinetic energy.

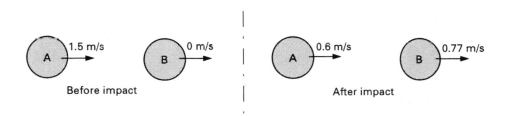

Figure 5.8 Solution to Example 4.6

We recall that $m_a = 0.3\,\text{kg}$ and $m_b = 0.35\,\text{kg}$.

Before impact

$$\sum E_k = \left(\tfrac{1}{2}mv^2\right)_a + \left(\tfrac{1}{2}mv^2\right)_b$$
$$\sum E_k = \left(\tfrac{1}{2} \times 0.3 \times 1.5^2\right) + \left(\tfrac{1}{2} \times 0.35 \times 0^2\right) = 0.3375\,\text{J}$$

After impact

$$\sum E_k = \left(\tfrac{1}{2} \times 0.3 \times 0.6^2\right) + \left(\tfrac{1}{2} \times 0.35 \times 0.77^2\right) = 0.054 + 0.104 = 0.158\,\text{J}$$

This suggests that the system has lost 0.1795 J. Not a tremendous amount, but when considering what was there to begin with, it is a loss of some 53%. This appears to break the principle of conservation of energy!

However, we have not really examined what has taken place. The objects collided, so some energy must have been used in the deformation that took place during the collision. Also, when they collided, they probably produced a bang, which required energy to drive it. Lastly the collision would have caused localised warming to occur, i.e. heat. Joule postulated that, in every event, energy ultimately transfers itself to heat.

This example does not teach us that our calculations are wrong but that our model of the collision is incomplete. We have not allowed for every possible event. A great many complex computer-aided analysis packages are based on the principles of conservation of momentum and conservation of energy. The more events they allow for (in energy terms, the more energies they allow for), the more accurate they become in modelling the real world. Hence a thorough understanding of these two principles is an essential item in an engineer's analytical toolbox.

Example 5.8   *A ball of mass* m *is dropped from the top of a building of height 20 m. Determine the velocity of the ball just before it hits the ground. Use the constant-acceleration equations then use the principle of conservation of energy. Confirm that the two methods produce the same answer.*

Constant-acceleration equation solution
First, let us sketch a picture of the problem, identifying all the boundary conditions (Figure E5.8).

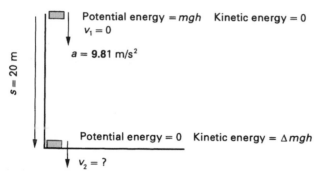

Potential energy $= mgh$ Kinetic energy $= 0$
$v_1 = 0$

$a = 9.81$ m/s^2

$s = 20$ m

Potential energy $= 0$ Kinetic energy $= \Delta mgh$

$v_2 = ?$

Figure E5.8

Recalling from Chapter 3 that final velocity, distance travelled and acceleration are linked by

$$v_2^2 = v_1^2 + 2as$$

we know that $v_1 = 0 : s = -20$ m and $a = -9.81$ m/s^2. This yields

$$v_2^2 = 2 \times (-9.81) \times (-20)$$

or

$$v_2 = -19.81 \text{ m/s}$$

Note: a square root yields two answers $+19.81$ and -19.81, but we pick -19.81 because we know the ball must be dropping. Your calculator does not tell you this, but it is worth remembering.

Conservation of energy solution
Assume there are no other forms of energy apart from potential and kinetic. All of the potential energy stored in the ball at the top of the building is 'lost' by the time it reaches the

bottom of the building. But it is not really 'lost', it is transferred to the motion of the ball as kinetic energy. Hence

change in potential energy = change in kinetic energy

$$\Delta E_g = \Delta E_k$$

assuming mass is constant, we have

$$mg(\Delta h) = \tfrac{1}{2}m(v_2^2 - v_1^2)$$

The boundary conditions are the same as in the previous solution, $\Delta h = 20\,\text{m}$ and $v_1 = 0$, so

$$mg(\Delta h) = \tfrac{1}{2}mv_2^2$$

We rearrange to yield an equation for v_2:

$$v_2 = \sqrt{2g\Delta h}$$

hence

$$v_2 = -19.81\,\text{m/s}$$

And this is identical to the previous solution. So why bother? The constant-acceleration equation solution will never change. For these conditions we can only ever achieve $v = 19.81\,\text{m/s}^2$. However, the energy solution can be easily modified to allow for other energy losses. The more types of energy we accommodate, the more accurate the model.

Example 5.9 *A round barrel of mass 45 kg and moment of inertia 12 kg m² is at the top of a hill (height 40 m). If the barrel's diameter is 400 mm, determine*

(a) the potential energy of the barrel at the top of the hill
(b) the linear velocity of the barrel at the bottom of the hill if it slides
(c) the linear velocity of the barrel at the bottom of the hill if it rolls

SOLUTION
(a) Figure E5.9(a) helps to establish the boundary conditions at A and B, leading to a solution. At the top of the hill the potential energy of the barrel, with respect to the base of the hill, is given by

$$\begin{aligned} E_p &= mgh \\ &= 45 \times 9.81 \times 40 \\ &= 17.658\,\text{kJ} \end{aligned}$$

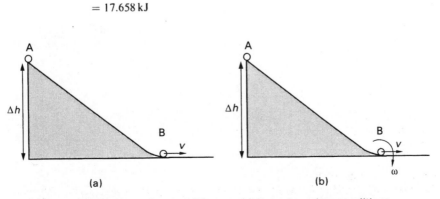

(a)   (b)

Figure E5.9 (a) Initial boundary conditions and (b) new boundary conditions

(b) We must now consider the change in energy of the barrel.

At A: $E_p = mg(h_1)$ and $E_k = 0$

At B: $E_p = mg(h_2)$ and $E_k = \frac{1}{2}mv^2$

The change in potential energy of the barrel is given by

$$\Delta E_p = mg(\Delta h)$$

From the principle of conservation of energy, the loss in potential energy that occurs as it slides down the slope must go somewhere. If we assume there is no other form of energy loss, all of this potential energy must change into kinetic energy of the barrel. Note that it is not the distance travelled that matters, but the height through which the barrel drops. Since the barrel was stationary at the top of the slope, we can write

$$\Delta E_p = E_k$$

or

$$mg(\Delta h) = \frac{1}{2}mv^2$$

which when rearranged becomes

$$v = \sqrt{2g\Delta h}$$
$$= \sqrt{2 \times 9.81 \times 40}$$

hence

$$v = 28.01 \text{ m/s}$$

(c) If we redraw the problem in Figure E5.9(b), we can recognise the new boundary conditions.

At A: $E_p = mg(h_1)$ and $E_k = 0$

At B: $E_p = mg(h_2)$ and $E_k = \frac{1}{2}mv^2 + \frac{1}{2}I\omega^2$

Using a similar analysis to part (a) we can see that the change in potential energy must be equal to the sum of both forms of kinetic energy (see Example 5.5). Hence we can write

$$mg(\Delta h) = \frac{1}{2}mv^2 + \frac{1}{2}I\omega^2$$

Recalling, from Chapter 2, that the peripheral speed of a body and its rotational speed are linked by $v = \omega r$, we can substitute an expression for ω (again see Example 5.5):

$$mg(\Delta h) = \frac{1}{2}mv^2 + \frac{1}{2}I(v/r)^2$$

or

$$mg(\Delta h) = \frac{1}{2}v^2(m + I/r^2)$$

which, when rearranged for v, becomes

$$v = \sqrt{\frac{2mg(\Delta h)}{(m + I/r^2)}}$$

thus

$$v = \sqrt{\frac{2 \times 45 \times 9.81 \times 40}{(45 + 12/0.2^2)}}$$

hence

$$v = 10.12 \, \text{m/s}$$

which is less than the answer to part (a). We would expect this since some of the energy has been used to cause the barrel to spin, and there is less for linear velocity to use.

Example 5.10 *A pinball machine fires small stainless steel ball bearings of mass 0.25 kg from a horizontal ejector mechanism. The ejector mechanism allows a spring, of stiffness 200 N/m, to be compressed by up to 45 mm. Once released, the stored energy of the spring accelerates the ball into the pinball machine.*

(a) Determine the eject velocity of the ball.

(b) Suggest an error in your model.

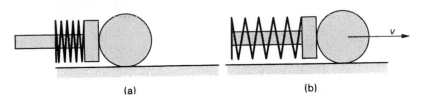

(a)   (b)

Figure E5.10   (a) Pinball mechanism fully compressed and (b) when the plunger and ball have just separated; the ball has its maximum velocity at this position

SOLUTION

Figure E5.10(a) illustrates the mechanism when fully compressed. The player does work on the spring to compress it. The spring stores this work in the form of elastic energy. Once released, the stored energy accelerates the mass and thus is converted into kinetic energy. The maximum velocity, and hence kinetic energy, occurs when the ball and plunger have separated, as shown in Figure E5.10(b). Recalling the equations for the stored energy in a spring and the kinetic energy of a moving body, we can write

$$\tfrac{1}{2}k\delta^2 = \tfrac{1}{2}mv^2$$

or

$$k\delta^2 = mv^2$$

which can be rearranged to yield an expression for v:

$$v^2 = \frac{k\delta^2}{m}$$

hence

$$v = \sqrt{\frac{200 \times 0.045^2}{0.25}}$$

or

$$v = 1.27 \, \text{m/s}$$

The errors which have slipped into this model are numerous. However, two stand out like proverbial sore thumbs:

(i)   We have neglected the friction between the ejector pin and its bearing.

(ii)   We have neglected the masses of the spring and the ejector.

Example 5.11   *The ejection system described in Example 5.10 is now fired vertically. Determine the maximum velocity attained by the projectile and the position where it occurs.*

SOLUTION

A sketch of the problem, Figure E5.11(a), illustrates the boundary conditions. Because the object rises, we must include the change in potential energy. To obtain a full picture of the energy transfer, we examine the energies at positions A, B and C.

At A all of the energy is stored elastic energy in the spring:

$$E_c = \tfrac{1}{2} k \delta^2$$

At B all of the energy has been converted to both kinetic energy and potential energy of the projectile:

$$\tfrac{1}{2} k \delta^2 = \tfrac{1}{2} m v^2 + mgh$$

At C there is a mix of energies; some stored energy is left in the spring, and the projectile itself has some potential energy and some kinetic energy. Using the compressed spring positon as a datum, we can formulate a model for the energies in the system:

$$\tfrac{1}{2} k \delta^2 = \tfrac{1}{2} m v^2 + mgh + \tfrac{1}{2} k (\delta - h)^2$$

where the term on the left-hand side is the original stored energy in the spring.

On the right-hand side, the first term is the potential energy of the ball at a height h, the second term is the kinetic energy of the ball at the height h, and the third term is the remaining stored energy in the spring. Figure E5.11(b) is a graph which represents the contribution of all the energies as height increases.

As can be seen in Figure E5.11(b), the kinetic energy reaches a peak at about 0.032–0.033 m, which is equivalent to a remaining spring compression of about 0.012–0.013 m.

The static deflection of the spring is given by

$$\delta_o = mg/k$$
$$= 0.25 \times 9.81/200$$
$$= 0.0123 \, \text{m}$$

which is very similar to the value described above.

Hence this example has shown that, if a spring is used as a vertical ejector, the maximum kinetic energy, and hence the maximum velocity, occurs at the point of static equilibrium.

This makes sense because, at static equilibrium, the body force generated by the mass of the object and the spring force are equal and opposite. When the spring is compressed beyond this point, the spring force exceeds the body force and the body accelerates upwards. Once the static equilibrium position is exceeded, the body force exceeds the spring force and the body is not accelerated by the spring. This is demonstrated in Figure E5.11(c).

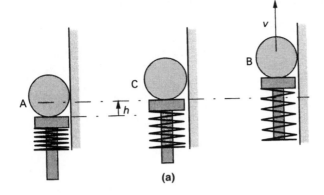

(a)

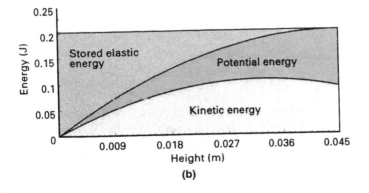

(b)

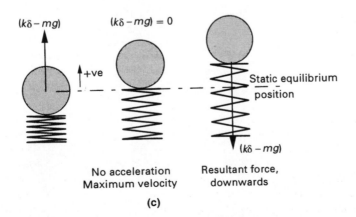

(c)

Figure E5.11 (a) Setting up the boundary conditions; (b) graph of energy against height of the ball; (c) once it passes the static equilibrium position, the ball is no longer accelerated by the spring

Applying our model

$$\tfrac{1}{2}k\delta^2 = \tfrac{1}{2}mv^2 + mgh + \tfrac{1}{2}k(\delta - h)^2$$

at $h = 0.0327\,\text{m}$ yields

$$\left(\tfrac{1}{2} \times 200 \times 0.045^2\right) = \left(\tfrac{1}{2} \times 0.25 \times v^2\right) + (0.25 \times 9.81 \times 0.0327)$$
$$+ \left[\tfrac{1}{2} \times 200 \times (0.045 - 0.0327)^2\right]$$
$$0.2025 = 0.125v^2 + 0.08 + 0.0151$$

hence

$$v^2 = 0.1074/0.125$$

thus

$$v = 0.927\,\text{m/s}$$

5.3 Power

Power is defined as

The rate of work, *or*, the rate of change of energy.

Power is given the symbol P and its unit is the watt (W).[4] By definition, rate of work may also be described as J/s.

To understand what power means, consider two motor cycles. One motor cycle is a small runabout, the other a full-blooded speed machine. The fuel tank stores the same fuel, petrol, and so has the same energy content. So what is the difference? Why is one motor cycle said to be more powerful than the other?

Although both motor cycles can store the same amount of energy, the more powerful machine can transform the stored energy in the fuel to kinetic energy of the motor cycle faster than the runabout is able to; it is said to have more power. Thus power is not a measure of energy, but of ability to transform energy.

Example 5.12 *Determine an expression for the power being transmitted by a shaft. The shaft is subject to a constant applied torque T and is rotating with angular velocity ω. Hence determine the output torque of a small lawnmower engine whose power output is 6 kW when operating at a speed of 3600 rev/min.*

SOLUTION
Recall that, for linear motion, work is defined as

$W = \text{Force} \times \text{distance}$

The equivalent for angular motion may be written

$W = \text{torque} \times \text{angle}$

---

[4] After James Watt, Scottish engineer and inventor, whose work on steam power was a key to the start of the Industrial Revolution (1736 1819).

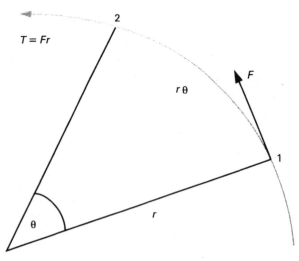

Figure E5.12

This is demonstrated in Figure E5.12. Notice that, if it moves from point 1 to point 2, the force F moves through an arc of $r\theta$. Hence the work done is given by

$$W = Fr\theta$$
$$= T\theta$$

Rate of change of work is defined by

$$P = \frac{dW}{dt}$$
$$= \frac{d(T\theta)}{dt}$$

but since T is constant

$$P = \frac{Td\theta}{dt}$$
$$= T\omega$$

In a similar fashion, the power of a constant force F moving at velocity v is given by

$$P = Fv$$

The output torque of the engine is therefore given by

$$T = \frac{P}{\omega}$$
$$= \frac{6000}{(3600 \times 2\pi)/60}$$
$$= 15.92\,\text{N m}$$

Summary

Definitions

Work is defined as the product of force and distance, where they are both in the same direction.

Work is also the area under the force–displacement graph (or equivalent). Mechanical work is carried out when bodies are in contact.

$W = Fx$ (for a constant force)

$W = \int Fdx$ (for a varying force)

Energy is the 'stored' potential to do work.
The unit of work and energy is the joule (J).
Power is defined as the rate of change of energy. Its unit is the watt (W) and $1\,W = 1\,J/s$

$P = \mathbf{F}\mathbf{v}$

$P = T\omega$

Energy forms

Potential energy: $E_p = mgh$

Kinetic energy, linear: $E_k = \frac{1}{2}mv^2$

Kinetic energy, angular: $E_k = \frac{1}{2}I\omega^2$

Stored elastic energy: $E_e = \frac{1}{2}k\delta^2$

Heat

Heat is the thermal equivalent of work.
Energy may be stored in a body in the form of heat.
Heat can be transferred without contact.
Heat is defined by

$Q = mc\Delta T$

Conservation of energy

Energy cannot be destroyed, but it can change from one form to another. Thus the total energy within a system remains constant.

Problems

Work and energy

5.1   A force of 100 N was applied to slide a small box over a distance of 2 m. Determine the work done by the force.

5.2   Table P5.2 illustrates the force displacement relationship for an automatic gate-opening mechanism. Draw a graph of force versus displacement and hence determine the work done to open the gate.

Table P5.2

Force (kN)	Displacement (m)
0	0
0.5	0.25
1.0	0.5
1.0	1
1.0	1.5
0.5	1.75
0	2.0

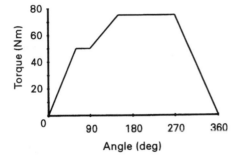

Figure P5.3

5.3   Figure P5.3 illustrates torque–angle data for one operation of a pressing machine. Determine the work done during the pressing process.

5.4   An aeroplane of mass 25 tonnes flies at a height of 1000 m and with a speed of 350 km/h. Determine the potential energy and the kinetic energy of the aeroplane.

5.5   A spring requires a force of 10 kN to compress it by 20 mm. Determine the stiffness of the spring and the elastic energy stored within the spring when it is compressed a further 12 mm.

5.6   A trolley of total mass 500 kg runs on four wheels of diameter 300 mm, each having a moment of inertia 1.5 kg m² and mass 12 kg. If the maximum speed of the trolley is 8 m/s, determine
   (a)   the angular velocity of the wheels
   (b)   the kinetic energy of the wheels
   (c)   the total kinetic energy of the trolley

5.7   (a)   Define the term *stiffness*.
   (b)   The results of an experiment to determine the stiffness of a spring are given in Table P5.7. A spring was mounted vertically on a frame and a succession of weights were hung from its base. Each time a weight was attached, the new length of the spring was measured; the applied mass and measured length are given in the table. From these results determine
       (i)   the stiffness of the spring
       (ii)   the energy stored in the spring when it has been extended by 45 mm

Table P5.7

Spring length (mm)	Applied mass (kg)
150	0
160	5
170	10
180	15
190	20
200	25
210	30

Conservation of energy

5.8   A drum of diameter 500 mm is at the top of a steep incline. The height of the incline, with respect to its base, is 25 m. If the drum slides down the slope, without rolling, determine the maximum velocity of the drum at the bottom of the slope.

5.9   If the drum, described in Problem 5.8, rolls down the hill, without slip, determine the linear and angular velocities of the drum at the base of the slope. The drum has a mass m of 25 kg and its moment of inertia I is $5.0 \, \text{kg m}^2$.

5.10   A projectile is fired vertically upwards from an ejection device with an initial velocity of 75 m/s. If the mass of the projectile is 15 kg, determine the maximum height reached by the object. Use the principle of conservation of energy to obtain your solution.

5.11   A ground-bait firing mechanism for anglers employs a spring of stiffness 500 N/m as its energy source. The mechanism compresses the spring by 50 mm and, when released, the stored energy in the spring fires a maximum of 0.25 kg of ground-bait out of a barrel. Determine
    (a)   the stored energy in the spring when compressed
    (b)   the maximum velocity attained when fired horizontally

5.12   A box of mass 2 kg is dropped on to a table of mass 15 kg from a height of 2 m. The table is supported by symmetrically mounted springs, giving a total stiffness of 2.5 kN/m. Determine
    (a)   the static deflection of the springs prior to the box being dropped
    (b)   the potential energy of the box relative to the table surface
    (c)   the maximum velocity the box achieves just before it hits the table
    (d)   the amount the table moves after the impact (you may assume that the box does not bounce)

5.13   An aircraft of mass 2 tonnes approaches the deck of an aircraft-carrier with a speed of 120 km/h. When landing, a hook at the rear of the aircraft catches on to a strong cable stretched across the deck. The cable itself is attached to springs of total stiffness 2 kN/m.
    (a)   Describe the energy transfer which occurs to stop the aircraft.
    (b)   Determine the amount the springs extend.
    (c)   State what happens to the aircraft once it comes to a halt.

5.14   A projectile of mass 2 kg collides with a stationary Plasticine model of mass 40 kg. The initial velocity of the projectile is 20 m/s and the impact is perfectly plastic.
    (a)   Determine the velocities of both objects after the collision.
    (b)   Determine the energy loss which occurred during the collision.
    (c)   State where the energy has gone.

5.15 A box is at the top of a steep frictionless slope of height 3.5 m. If the box is allowed to slide down the slope, determine its maximum velocity at the base of the slope.

5.16 A cylinder of mass 10 kg, moment of inertia 5 kg m^2 and diameter 1 m rolls freely with a linear velocity of 2.5 m/s; determine its total kinetic energy.

5.17 A ski-jumper stands at the top of a specially prepared slope. The skier accelerates down the slope, which gradually adjusts the direction of travel of the skier from a 'downward' trajectory to one which is horizontal.

   (a) If the vertical distance from the top of the slope to its base is 80 m, determine the maximum velocity of the skier at its base.

   (b) If the end of the ski-jump is 10 m above ground level, which is perfectly flat, determine the nominal distance of a ski-jump.

   (c) How can this distance be improved?

Power

5.18 Show that the rate of work, power, of a force F moving with velocity v is $P = Fv$.

5.19 The power output of a large domestic lawnmower engine is 9 kW at an engine speed of 3600 rev/min. Determine the output torque of the engine.

5.20 A jet engine produces a thrust of 60 kN when the aircraft flies at a nominal speed of 1000 km/h. Determine the minimum power required to achieve this.

5.21 (a) If the average torque of a process is defined by

$$T_{ave} = \frac{\text{work done by process}}{\text{angle covered by process}}$$

   determine the average torque for the process described in Problem 5.3.

   (b) If the process takes 12 s to complete, determine the average power required.

Chapter 6

Modelling using free-body diagrams

In previous chapters we met physical laws which allow us to model real-life situations mathematically. Another tool which is essential to the engineer is a conceptual one. It is the art of drawing *free-body diagrams*. Because it is conceptual, students often find it to be a demanding topic and difficult to understand fully. But, as with most conceptual ideas, it is practice which makes perfect. This chapter will introduce the basic concepts then expand upon them to give a fuller understanding.

6.1   Definition of a free-body diagram

A free-body diagram is a diagrammatic representaion of a body which has been separated from *all other* bodies and which is drawn as if it were floating in free space. All forces and moments which act *on* the body are then drawn such that they are represented by arrows positioned at the point of *actual* application.

Consider the case of a heavy ball suspended from a ceiling by a rope, as illustrated in Figure 6.1. Let us now imagine the ball as if it were floating in space, with no

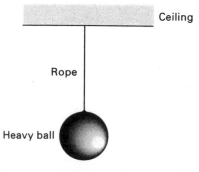

Figure 6.1   A heavy ball suspended from the ceiling by a rope

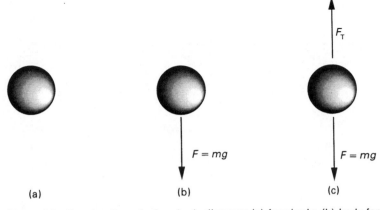

Figure 6.2 Construction of a free-body diagram: (a) free body, (b) body force attached, (c) other forces added

forces acting on it. That is our free body (depicted in Figure 6.2(a)). The next step, the engineering part of the task, is to imagine all the forces that are acting on the ball; this part takes practice and experience. But in this situation you should be able to see that the ball must be subject to a force due to gravity:

$$\mathbf{F} = m\mathbf{a}$$

where $\mathbf{a} = \mathbf{g}$ and hence

$$\mathbf{F} = m\mathbf{g}$$

which we know (from Chapters 3 and 4) acts straight down. The force induced by gravitational acceleration is called the *body force*. It exists *always*. The body force acts through the body's *centre of gravity*.

The centre of gravity of an object can best be decribed by considering any arbitrary object hung from a point. In Figure 6.3(a) the object is hung from point A, and the line drawn from this point vertically downwards is the line through which the body force acts. If we now repeat this process for points B and C we obtain several load-lines, as demonstrated in Figure 6.3(b) and (c), but they all cross at the same point. This point is the body's centre of gravity and may also be considered as the point about which the body balances. When drawn on a figure or diagram, the centre of

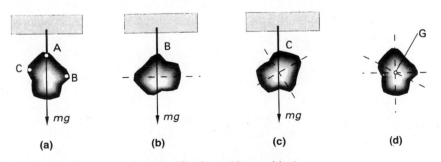

Figure 6.3 The centre of gravity (G) of an arbitrary object

gravity is always donated by G. We can now add the body force to the diagram (Figure 6.2(b)).

We also recognise that attached to the ball there is a rope, which is probably exerting some force on the ball. Ropes, cables and strings can only pull – you cannot push with a rope – hence forces induced by cables and ropes are called *tension* (after tensile forces) and we give them the symbol F_T. The tension in the rope will act straight up; although we do not know its value, we can add the tension to the free-body diagram (Figure 6.2(c)).

We could also draw a free-body diagram for the point where the supporting cable is attached to the ceiling. Here we would obtain the cable tension acting down (because it is transmitting the 'weight' of the ball), and the ceiling opposing this downward force, i.e. a force vertically upwards. The opposition of a force by a wall, ceiling, floor, support or any other object is called a *reaction*; it is often given the symbol **R**.

Note that the forces are always drawn where they act. Also, because the arrows indicate direction, only the magnitude of the force is used for annotation, no sign is required! Once all the arrows have been drawn, we can use Newton's second law to solve for any unknowns and to determine resultant forces acting on bodies.

Before any further examples can be made, four special cases have to be considered.

6.1.1   Statically determinate systems

Statically determinate systems are said to be in *static equilibrium*. Hence we can use Newton's second law applied to static bodies to determine *all* unknown forces and moments:

$$\Sigma \mathbf{F} = 0$$

and

$$\Sigma \mathbf{M} = 0$$

Implicit within this statement is that a system may be reduced to x and y components for simplification.

6.1.2   Statically indeterminate systems

As in Section 6.1.1, the system is also in static equilibrium, but in this case the use of Newton's second law does not solve all unknown equations. Hence the application of *statics* principles is not sufficient and we must look at other boundary conditions. Indeterminate systems are beyond the scope of this text but will undoubtedly be met in later studies.

6.1.3   Concurrent forces

In this case all forces point to, or away from, the same place. They are *coincident* at one point. Figure 6.4(a) illustrates concurrent forces which act in the x–y plane only. Figure 6.4(b) illustrates 3D concurrent forces. When forces are concurrent, equilibrium is examined using $\Sigma \mathbf{F} = 0$ only.

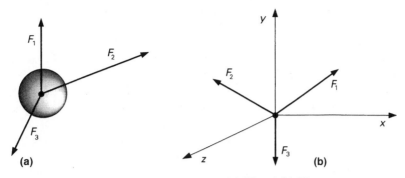

Figure 6.4 Examples of concurrent forces: (a) 2D and (b) 3D

6.1.4 Coplanar forces
In this case all forces are two-dimensional vectors acting in the same plane, either x–y, y–z, or z–x. Coplanar forces need not be concurrent. Figure 6.5(a) illustrates coplanar forces which are also concurrent. Figure 6.5(b) illustrates coplanar forces which are not concurrent.

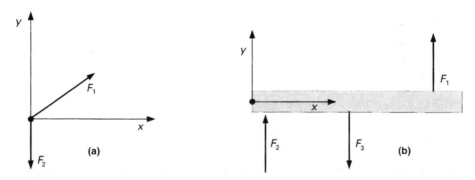

Figure 6.5 Examples of coplanar forces: (a) concurrent forces, (b) non-concurrent forces

When forces are coplanar and are not concurrent then both $\Sigma \mathbf{F} = 0$ and $\Sigma \mathbf{M} = 0$ need to be applied to examine system equilibrium.

6.2 Application to static systems

6.2.1 Triangle of forces
The *triangle of forces* is a useful tool for modelling a body which is subject to three coplanar forces:

> *If three forces, which are not parallel, are in static equlibrium then they must also be concurrent. Furthermore, their magnitude and direction are represented by a closed triangle.*

This is sometimes called *Lami's theorem*.

Example 6.1   *A winch is supported by two cables, as illustrated in Figure E6.1(a). The winch itself is supporting a mass of 500 kg. Determine the tension in each cable if the mass of the winch may be considered negligible.*

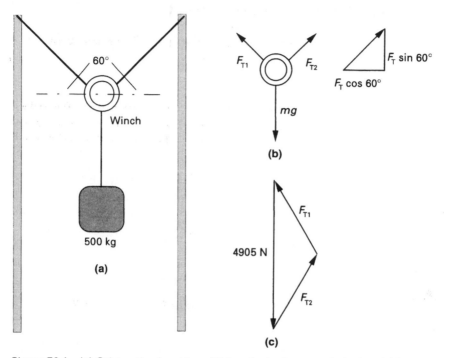

Figure E6.1   (a) Schematic of problem, (b) free-body diagram of winch and (c) triangle of forces

SOLUTION
First draw a free-body diagram of the winch and orient $+y$ vertically upwards, as shown in Figure E6.1(b).

STANDARD SOLUTION
By inspection we note the symmetry of the system, hence the magnitudes

$$F_{T1} = F_{T2}(=F_T)$$

Applying Newton's second law (in the y-direction only), $\Sigma F_y = 0$, hence

$$-(500 \times 9.81) + (2F_{Ty}) = 0$$
$$F_{Ty} = 4905/2 = 2.45\,\text{kN}$$

We know that

$$F_{Ty} = F_T \sin 60°$$

hence

$$F_T = F_{Ty}/\sin 60° = 2.83\,\text{kN}$$

Using the triangle of forces we can draw the force vectors to scale and achieve Figure E6.1(c). We may also use the trigonometric functions given in Appendix A to obtain the magnitude of F_T.

6.2.2   Levers and mechanical advantage

Archimedes first studied levers in the third century BC, and he is famous for having said:

> *Give me somwhere to stand, and I shall move the earth.*

The principle of levers is used intuitively by almost every living person, and the analysis of how they work gives an insight into statics.

Example 6.2   *Consider the rigid beam depicted in Figure 6.6, which is a model of a playground see-saw. On each seat sits a child of mass 17 kg. Determine the reaction at the support and show that the see-saw is in static equilibrium.*

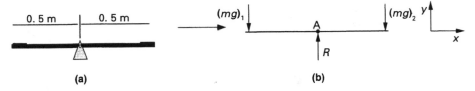

(a)   **(b)**

Figure 6.6   Reducing a model of a see-saw to a free-body diagram (forces only):
(a) schematic of beam and (b) free-body diagram

SOLUTION
The see-saw is reduced to its free-body diagram in Figure 6.6(b). The pivot, or *fulcrum*, is depicted as a single upward force and the *weights* of the children are depicted as the downward forces $(mg)_1$ and $(mg)_2$.

We should recognise this system as a coplanar/non-concurrent force system. Thus to check equilibrium we apply both $\Sigma F = 0$ and $\Sigma M = 0$.

To simplify the solution we orient the positive y-axis vertically upwards. Applying Newton's second law reveals that for equilibrium

$$\Sigma F_y = 0$$

or

$$-(17 \times 9.81) - (17 \times 9.81) + R = 0$$

hence

$$R = 2 \times 166.77 = 333.54\,\text{N}$$

Taking moments about the point at A (see Figure 6.7) and applying $\Sigma M_z = 0$ yields

$$-(0.5 \times 17 \times 9.81) + (0.5 \times 17 \times 9.81) = 0$$

Hence the see-saw is in static equilibrium since $\Sigma F = 0$ and $\Sigma M = 0$.

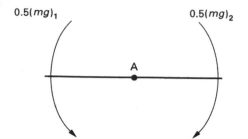

$0.5(mg)_1$ $0.5(mg)_2$

A

Figure 6.7   Free-body diagram of see-saw (moments only)

Example 6.3   *A belt tensioner on a ride-on lawnmower operates as illustrated in Figure E6.3(a). The cable may be tensioned at A with a thumbscrew, this in turn applies a force at B via the tension arm. Determine the mechanical advantage of the tension arm.*

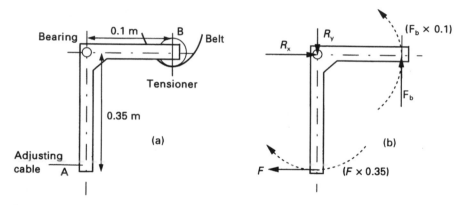

Figure E6.3   (a) Belt tensioner and (b) its free-body diagram

SOLUTION
Draw a free-body diagram of the applied forces (Figure E6.3(b) with the *y*-axis vertical. The forces are separated from the bearing, and hence the point of rotation. So there are resulting moments about the bearing. These have been added to the figure for clarity only. Only forces are applied, so only they should be drawn. The moments are a consequence of the forces.

Applying Newton's second law to the applied forces shows that in both the *x* and *y* directions the forces are in equilibrium:

$$\Sigma F_y = F_b - R_y = 0$$
$$\Sigma F_x = R_x - F = 0$$

Taking moments about the bearing reveals that for equilibrium

$$\Sigma M = 0$$
$$(F_b \times 0.1) - (F \times 0.35) = 0$$

or

$$0.1 F_b = 0.35 F$$

hence

$$\frac{F_b}{F} = \frac{0.35}{0.1} = 3.5$$

This value tells us that the force applied at the tensioner is 3.5 times greater than the force applied by the tension cable. This is called *mechanical advantage*.

6.2.2 Systems subject to friction

In practice, friction is a force which always exists when the surfaces of two bodies are in mutual contact. Friction may be modelled very simply using

$$F_f = \mu R_n \qquad (6.1)$$

where

F_f = the frictional force which always opposes motion

μ = the coefficient of friction between the two surfaces

R_n = the normal reaction force acting on the contracting surfaces

Note that equation (6.1) is independent of area, a fact which is not often appreciated. Figure 6.8 illustrates the terms used in equation (6.1). Note that the frictional force is drawn as opposing the motion and it always acts *at the surface*.

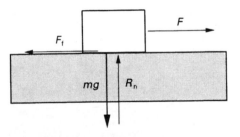

Figure 6.8 A body subject to a frictional force F_f

It is beneficial to think of the frictional force as a limiter. The force determined by equation (6.1) is the *maximum* value which can be obtained. In fact any two bodies which are in contact experience a frictional force that is just sufficient to oppose motion - equilibrium is maintained. Once the driving force exceeds the value given by equation (6.1), then sliding begins. If this were not the case, you would find pens, cups and books sliding across tables; friction would be inducing motion! Friction always *opposes motion*.

There are two models for the coefficient of friction, static and dynamic. The static friction coefficient μ_s applies prior to one body sliding on the other. This coefficient is highly dependent on the materials which are in contact, their condition, any impurities, lubrication or debris between the surfaces and even on the atmosphere surrounding them. Quick calculations often assume a value of $\mu_s = 0.3$, but British Standards and many engineering handbooks give tables of specific values for specific materials.

Once sliding has begun, the static friction coefficient is inapplicable and the coefficient to be used is the dynamic friction coefficient μ_k. This coefficient is a constant, unless the surface velocity is either very low or very high.

Example 6.4   *A box containing engine components has a combined mass of 450 kg. The box is in a warehouse where the floor is smooth concrete. If it takes a force of 1.1 kN to move the box determine the coefficient of friction between the floor and the box.*

SOLUTION
Draw a free-body diagram of the box (Figure E6.4). Note that the external force is assumed to be applied at the base.

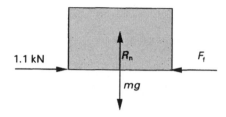

Figure E6.4

Just before the box starts to move, the force applied is 1.1 kN. Hence, from Newton's second law, the maximum frictional force opposing motion (F_f) must be 1.1 kN. The body force of the box is

$$mg = 450 \times 9.81 = 4414.5\,N$$

which is opposed by the reaction of the floor, hence the normal reaction force is

$$R_n = 4414.5\,N$$

Using equation (6.1)

$$F_f = \mu R_n$$
$$1100 = \mu \times 4414.5$$

hence

$$\mu = 1100/4414.5 = 0.25$$

Example 6.5   *A thin, wide plate lies on a flat surface which can be tilted to any angle about a hinge at one end (as illustrated in Figure E6.5(a)). If the static coefficient of friction between the plate and the table surface is $\mu_s = 0.29$, determine the angle at which the plate will begin to slide.*

SOLUTION
Since the object is going to slide down the slope, it makes sense to orient our x and y axes with the table, not with what we perceive as vertical, as shown in Figure E6.5(b). Thus we can take components of the body force (mg) which act down the slope and normal to the table surface (Figure E6.5(c)). Now we can draw a free-body diagram for the plate, adding the

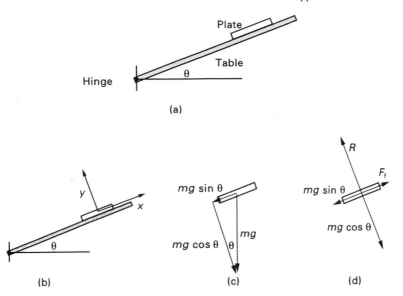

Figure E6.5   (a) Schematic, (b) coordinate system, (c) body-force components and
(d) free-body diagram

reaction force of the table surface upon the plate (Figure E6.5(d)). Since the plate is thin, we
can treat the position of the centre of gravity G and the contact surface as being the same.

Using the free-body diagram we can determine our unknown forces F_f and R_n. From
Newton's second law (statics) we know that

$$\Sigma F_x = 0$$

An examination of the free-body diagram in the x-direction reveals that

$$F_f - mg \sin \theta = 0$$

or

$$F_f = mg \sin \theta \tag{1}$$

A.so $\Sigma F_y = 0$ hence

$$R_n - mg \cos \theta = 0$$

or

$$R_n = mg \cos \theta \tag{2}$$

From equation (6.1) we also know that

$$F_f = \mu_s R_n$$

into which we can substitute (1) and (2), hence

$$mg \sin \theta = \mu_s mg \cos \theta$$

which when rearranged becomes

$$\mu_s = \frac{\sin \theta}{\cos \theta} = \tan \theta$$

or

$$\mu_s = \tan \theta \tag{3}$$

hence the angle at which the plate will begin to slip is given by

$$\theta = \tan^{-1} \mu_s$$
$$= \tan^{-1} 0.29$$
$$= 16.17°$$

Note that equation (3) is a general solution for any object on an inclined surface. But if the object's thickness becomes too large, this equation breaks down because the object can topple as well as slide, which we have not allowed for! Try this experiment using several different objects and determine values of μ_s for yourself.

Example 6.6   *A block (of mass 500 kg) leans against a 300 mm high step, as illustrated in Figure E6.6(a). If the support at A is frictionless, determine the reaction forces at A and B. Hence determine the friction coefficient between the block and the floor at B.*

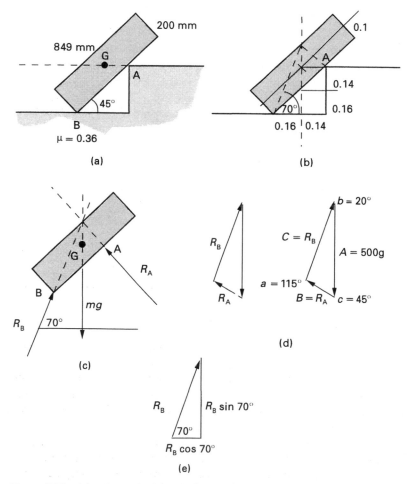

Figure E6.6   (a) schematic, (b) calculated dimensions, (c) free-body diagram, (d) triangle of forces and (e) reaction force at B

SOLUTION

The geometry of the system is illustrated in Figure E6.6(b). Drawing a free-body diagram for the block highlights the direction of the unknown reaction forces at A and B (Figure E6.6(c)). Note that the force at A must be normal to the surface of the block because there is no friction. The angle of the reaction force at B can therefore be obtained by drawing the block and step to scale or, preferably, by geometry.

The body force acts through G, we know the direction of the reaction force at A and they must cross as indicated. The reaction force at B must also cross this point (Lami's theorum).

Using the principle of the triangle of forces, we can construct a closed force diagram as illustrated in Figure E6.6(c). Hence we can find the magnitudes of R_A and R_B from a scale drawing or by geometry.

Using Appendix A, we note that the lengths of the sides of any triangle and its included angles are given by

$$B = \frac{A \sin b}{\sin a} \quad \text{and} \quad C = \frac{A \sin c}{\sin b}$$

Assigning $A = 500\,\text{g}$, $a = 115°$, $b = 20°$ and $c = 45°$, as illustrated in Figure E6.6(d), we obtain

$$R_B = (500 \times 9.81)\frac{\sin 45°}{\sin 115°} = 3827\,\text{N at } 70°$$

$$R_A = (500 \times 9.81)\frac{\sin 20°}{\sin 115°} = 1851\,\text{N at } 225°$$

The force R_B has a vertical component, due to the reaction against the body force, and a horizontal component, due to friction. This is illustrated by the triangle of forces in Figure E6.6(e). Since

$$F_f = \mu R_n$$
$$R_B \cos 70° = \mu R_B \sin 70°$$

thus

$$\mu = \cos 70°/\sin 70°$$
$$= 0.36$$

Which is also tan 20°. Hence, whenever there is friction at a surface, the direction of the reaction force is always known because $\theta = \tan^{-1} \mu = (90° - \text{reaction-force angle})$.

6.2.4 *Pulley-rope systems*

Pulleys and ropes have been in use for hundreds of years (nowadays we also use cables). The reason for their popularity is quite simply ease of use and that a tremendous amount of mechanical advantage is readily achieved.

A rope or a cable can only apply a force in one direction, along its length. A rope or a cable can only be pulled; you cannot expect to achieve much by pushing a rope! This is exemplified in Figure 6.9. In engineering terms it means that a rope or cable can only transmit tensile not compressive forces.

Pulleys enable ropes to 'go round corners', as illustrated in Figure 6.10. They are discs which run on bearings, and the rope or cable sits in a machined groove on the periphery. If the interaction between the rope and the pulley is assumed to be frictionless, the tension within the rope on one side of the pulley is identical to that on the other side.

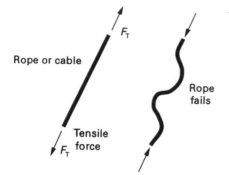

Rope or cable

F_T

Rope
fails

Tensile
force
F_T

Figure 6.9   Transmission of force by a cable or rope

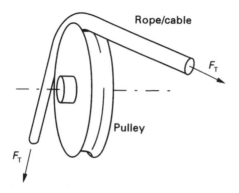

Rope/cable

F_T

Pulley

F_T

Figure 6.10   A rope transmitting force around a pulley

Example 6.7   *A tree surgeon, of mass 76 kg, is lowering equipment down the side of a cliff in order to prune a large tree halfway down the cliff face. The equipment, of combined mass 12 kg, is tied to a rope which itself runs on a small pulley at the top of the cliff and over a branch of a nearby tree so that the rope can be pulled vertically, as shown in Figure E6.7(a). Determine the tension in the rope and the 'apparent mass' of the tree surgeon.*

SOLUTION

Drawing a free-body diagram of the package (Figure E6.7(b)), reveals that $F_T = 117.7$N, which is the tension in the rope. If we assume the pulley to be frictionless, F_T is constant along the rope's entire length. Thus we can draw a free-body diagram for the person (Figure E6.7(c)). Applying Newton's second law in the y-direction, which we orient vertically for convenience, gives

$$\Sigma F_y = 0$$
$$-(76 \times 9.81) + 117.7 + R = 0$$

hence

$$R = 627.86 \text{ N}$$

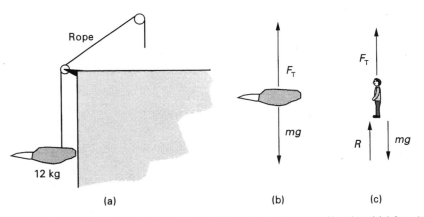

Figure E6.7 (a) Schematic of problem, (b) free-body diagram of load and (c) free-body diagram of person

The apparent mass of the person is what an imaginary set of bathroom scales would measure if the tree surgeon were standing on them:

$$m = \frac{R}{g}$$

$$m = \frac{627.86}{9.81} = 64 \, \text{kg}$$

Example 6.8 *A marine engine is being hoisted into a yacht. The engine's mass is 120 kg. The fitter constructs a lifting-hoist which consists of two pulley blocks (often called a block and tackle); one attached to the engine and one attached to a gantry. A rope is wound round the blocks such that it makes four complete revolutions of the blocks (as shown in Figure E6.8(a)). Determine the tension in the rope and the mechanical advantage of the system.*

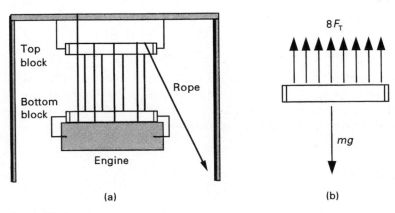

Figure E6.8

SOLUTION

We recognise that all the forces are parallel. This can be simplified further by recognising that the tension in the rope is the same throughout, hence all the forces are identical. To determine the tension in the rope we orient the y-axis vertically, for simplicity, and apply Newton's second law to the bottom block:

$$\Sigma F_y = 0$$
$$-(120 \times 9.81) + 8F_T = 0$$

hence

$$F_T = 147.15 \, \text{N}$$

The mechanical advantage of the system is 8, since an effort of 147.15 N supports a load of 1177.2 N.

6.2.5 Modelling concurrent forces

Consider the coplanar free-body diagram illustrated in Figure 6.11. If all the forces are referenced to the x-axis (as illustrated), the resultant force acting on the body is easily determined.

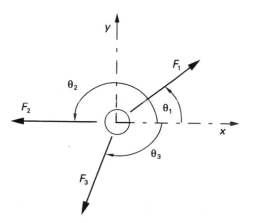

Figure 6.11   Three concurrent forces acting on a body

From Chapter 1, the resultant vector acting on the body is determined by vector addition. This can be done by drawing a vector diagram or, preferably, by adding the x and y components of each force individually.

Thus the x-component of the resultant force is given by

$$F_x = F_1 \cos \theta_1 + F_2 \cos \theta_2 + F_3 \cos \theta_3 + \dots$$

and the y-component is

$$F_y = F_1 \sin \theta_1 + F_2 \sin \theta_2 + F_3 \sin \theta_3 + \dots$$

The resultant force $\mathbf{F}$ is then determined using equations (1.1) and (1.2).

Example 6.9 *A joint in a seating stand is subject to the forces illustrated in Figure E6.9. Show that the joint is in static equilibrium.*

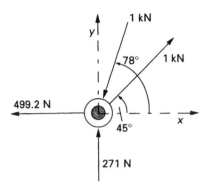

Figure E6.9

SOLUTION

First, list the known forces, using the x-axis as the reference datum:

$\mathbf{F}_1 = 1000$ at $45°$

$\mathbf{F}_2 = 1000$ at $-102°$ (note the direction of this force)

$\mathbf{F}_3 = 499.2$ at $180°$

$\mathbf{F}_4 = 271$ at $90°$ (note the direction of this force too)

Hence

$$F_x = 1000 \cos 45° + 1000 \cos - 102° + 499.2 \cos 180° + 271 \cos 90° = 0\,\text{N}$$

and

$$F_y = 1000 \sin 45° + 1000 \sin - 102° + 499.2 \sin 180° + 271 \sin 90° = 0\,\text{N}$$

Therefore the system is in static equilibrium.

6.3 Application to dynamic systems

In this case we are considering the *kinetics* of the system. This is the study of the motion of a body due to applied forces and/or moments. Here we apply Newton's second law in its dynamic form:

$$\Sigma\mathbf{F} = m\mathbf{a}$$

and

$$\Sigma\mathbf{M} = I\alpha$$

On a free-body diagram for dynamic systems there is another force, or moment, to be drawn. This is the *inertial force* of the body which resists acceleration, or which resists change of momentum.

The creation of an imaginary *dynamic equilibrium* using an inertial force ($\Sigma\mathbf{F} - m\mathbf{a} = 0$) is attributed to d'Alembert[1] and hence this is often termed d'Alembert's principle. Figure 6.10 illustrates how the inertial force is added to a free-body diagram. Note that it is commonly an unknown quantity and should always be aligned with the positive x, y or z axes. Sometimes it also pays dividends to make the inertial force bold to remind you that it is normally an *unknown*.

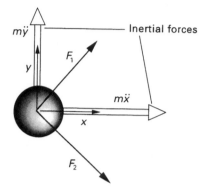

Figure 6.12   Free-body diagram with inertial force

Figure 6.12 illustrates another form of shorthand. Acceleration and velocity can be in any direction, but when they are aligned with x, y or z axes then dot notation can be used (see Chapter 2):

$$\dot{x} = \frac{dx}{dt} = v_x = \text{velocity in the } x\text{-direction}$$

$$\ddot{x} = \frac{d\dot{x}}{dt} = a_x = \text{acceleration in the } x\text{-direction}$$

similarly for the y-axis ($\dot{y}$ and $\ddot{y}$), and rotation θ (angular velocity $= \dot{\theta}$ and angular acceleration $= \ddot{\theta}$). This shorthand becomes more useful when complex models are to be developed, where motion may be in more than one axis. Such equations can become large and cumbersome. The dot notation helps to simplify them.

Example 6.10   *In a new enterprise initiative, weighing-machines have been fitted in all lifts in a department store (Figure E6.10(a)). However, customers are complaining about their accuracy. One person, in particular, states that, when travelling from the basement to the fifth floor, the machine said that his mass was 85 kg, although he knew it should be around 75 kg.*

(i) The lift accelerates at 1.3 m/s²; confirm that this acceleration is the source of the error.

(ii) If the total mass of the lift is 530 kg, determine the force required to accelerate the lift by 1.3 m/s².

---

[1] Jean Le Rond d'Alembert, French mathematician and physicist (1717–1783).

SOLUTION

(i) The free-body diagram is illustrated in Figure E6.10(b)

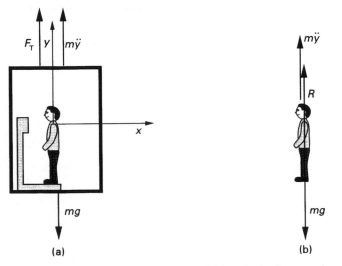

Figure E6.10 (a) Schematic of lift and (b) free-body diagram of person

Note that R exists because of contact between the person's feet and the scales. Applying Newton's second law to the person

$$\Sigma F_y = m\ddot{y}$$
$$-(75 \times 9.81) + R = (75 \times 1.3)$$

hence

$$R = 75 \times (9.81 + 1.3) = 833.25 \, \text{N}$$

The reaction force is measured by the scales which are calibrated using $a = 9.81 \, \text{m/s}^2$, hence they state that the person's apparent mass is

$$m = 833.25/9.81 = 84.94 \, \text{kg}$$

which confirms the acceleration as the source of the error.

(ii) The force required to produce this acceleration may be obtained from the free-body diagram of the lift. Applying Newton's second law

$$\Sigma F_y = m\ddot{y}$$
$$F_T - (530 \times 9.81) = (530 \times 1.3)$$

hence

$$F_T = 530 \times (9.81 + 1.3) = 5888.3 \, \text{N}$$

which is induced by a cable attached to the roof of the lift body. Modern lifts that do not have to travel to many floors are sometimes powered by hydraulic rams under the lift body, but the same force would still apply.

Example 6.11 *A funfair ride models an aeroplane in freefall, which is where an object drops under gravitational acceleration. Determine the apparent mass of the occupants of this ride.*

SOLUTION
Drawing a free-body diagram of an occupant (Figure E6.11) and applying Newton's second law, we have

$$\Sigma F_y = m\ddot{y}$$
$$-(m \times 9.81) + R = -(m \times 9.81)$$

y

$m\ddot{y}$

R

mg

Figure E6.11

hence

$$R = 0$$

thus the occupants feel as if they are weightless.

Summary

Free-body diagrams
There are two main classifications of static systems:

Statically determinate: all forces and moments sum to zero, $\Sigma F = 0$ and $\Sigma M = 0$. These conditions are sufficient to determine all unknown forces.

Statically indeterminate: all forces and moments sum to zero, $\Sigma F = 0$ and $\Sigma M = 0$. These conditions are not sufficient to determine all unknown forces.

Triangle of forces
If three non-parallel forces are in equilibrium, they are also concurrent. Their magnitudes can be estimated from a closed triangle.

Reaction forces
Whenever there is contact between two bodies there is always a reaction force.

Pulleys and ropes

If a pulley is assumed frictionless, the tension in the rope is the same throughout its length.

Dynamic systems

$$\Sigma F = ma$$

and

$$\Sigma M = I\alpha$$

Problems

In all cases a free-body diagram should be drawn or sketched

Static systems

6.1 State how Newton's second law of motion is applied to statics.

6.2 Three forces act on a pin-joint as indicated below:

$$F_1 = 15\,kN \text{ at } 45^\circ$$
$$F_2 = 7\,kN \text{ at } 175^\circ$$
$$F_3 = 5\,kN \text{ at } 270^\circ$$

(a) Determine the horizontal component of the resultant force acting on the joint.

(b) Determine the vertical component of the resultant force acting on the joint.

(c) Determine the resultant force acting on the joint.

(d) If the system is truly static what is required to achieve this?

6.3 A support, which is fixed to a wall, carries three loads as given below. Determine the reaction force produced by the wall.

$$F_1 = 15\,kN \text{ at } 210^\circ$$
$$F_2 = 7\,kN \text{ at } 180^\circ$$
$$F_3 = 18\,kN \text{ at } -20^\circ$$

6.4 A ladder, 2.5 m long, leans against a frictionless wall, as illustrated in Figure P6.4. If the mass of the ladder is 25 kg and its centre of gravity is at its geometric centre, determine the reactions at the wall and the floor. Also determine the coefficient of friction between the ladder and the floor.

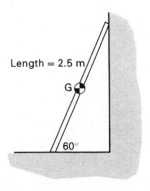

Length = 2.5 m

G

60°

Figure P6.4

6.5   If for Problem 6.4 there is a coefficient of friction of 0.3 between the ladder and the floor and between the ladder and the wall, determine the reactions at the wall and the floor.

6.6   Figure P6.6 illustrates an engine block being supported by a single rope hung from a pulley, which is rigidly fixed to the ceiling. The free end of the rope is tied to a fixing in a wall. Draw free-body diagrams of the engine, the pulley block and the wall fixing, hence determine the tension in the rope, the reaction force at the ceiling and the reaction force at the wall fixing.

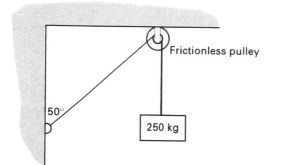

Figure P6.6

6.7   A lift, of mass 125 kg, is supported by seven cables attached to the centre of the roof, as illustrated in Figure P6.7. To stop 'twisting', the lift also runs on frictionless rails, as illustrated in the figure.
(a)   A passenger of mass 85 kg enters the lift and stands in the exact centre. Determine the tension in each of the cables.
(b)   The passenger moves to the left-hand side of the lift and the new centre of gravity of the lift is 0.25 m to the left of the original centre of gravity. Determine the tension in the cables and the reaction forces at the runners.

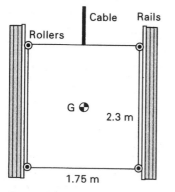

Figure P6.7

6.8   A car of mass 0.75 tonnes is parked on a hill such that the side of the car faces downhill. The centre of gravity of the car is positioned in the exact middle of the wheels and 0.5 m above ground level. The wheels are separated, across the car, by a distance of 2 m. If the slope of the hill is 10˚, determine the reaction forces on the wheels.

6.9   A car of mass 850 kg negotiates a bend in the road, of radius 250 m, at a speed of 75 km/h. The centre of gravity of the car is at the exact centre between the wheels, and the wheels are separated by 1.75 m across the width of the car. The centre of gravity of the car is also 0.4 m above ground level.

(a)   Determine the reaction forces on the wheels when the car is stationary.

(b)   Including body and centrifugal forces, draw a free-body diagram of the car when it negotiates the bend.

(c)   Determine the reaction forces at the wheels when the car negotiates the bend.

(d)   Determine the minimum coefficient of friction to avoid slip between the tyres and the road.

6.10   (a)   The framework for seating in an auditorium is fixed by a joint permanently attached to a solid wall. The forces produced by the seating which act on the joint may be modelled by three forces:

$$F_1 = 15 \text{ kN at } 210^{\circ}$$
$$F_2 = 10 \text{ kN at } 180^{\circ}$$
$$F_3 = 12 \text{ kN at } -20^{\circ}$$

Determine the reaction force produced by the wall on the fixing.

(b)   The permanent fixing is to be replaced with a friction pad. The pad is to be attached to the seating but is 'pressed' against the wall by the forces induced by the seating. Determine the minimum value of the coefficient of friction between the pad and the wall to enable this new fixing to function. Discuss the viability of this option.

6.11   A block of mass 250 kg rests on a flat surface. If the surface can tilt about one end and the coefficient of friction between the table and the block is 0.4, determine from first principles the angle at which the block begins to slide.

6.12   A force of 500 N acts on one end of a lever, 250 mm long, and its other end is fixed. Determine the reactions at the fixed end.

Dynamic systems

6.13   State how Newton's second law of motion is applied to dynamics.

6.14   An orbital space station has been constructed to form a large torus of internal diameter 200 m. The torus itself is constructed from hollow tube of diameter 15 m. The station rotates about its centre so as to produce a 'false gravity' of 1 g (9.81 m/s^2). Describe how this false gravity may be produced and hence determine the speed at which the station should rotate.

6.15   A sledge and its rider are at the top of a steep hill. The hill's slope is constant and its height changes by −3 m over a 2 m run. The total height of the hill is 25 m.

(a)   If the slope is considered frictionless, determine the maximum velocity at the bottom of the slope.

(b)   If the dynamic coefficient of friction between the sledge and the snow is 0.1, determine the maximum velocity at the base of the hill.

6.16   Heavy particles which are suspended in a fluid are often separated in a *centrifuge*. Research how a centrifuge operates then deduce, using free-body diagrams, how the solids are separated from the fluid. You may wish to consider the operation of a spin-drier!

6.17   A flywheel of moment of inertia 12.5 kg m^2 is driven by a motor whose maximum torque is 30 N m. The flywheel is also subject to a braking force of 30 N acting at its periphery. If the diameter of the flywheel is 1.5 m, determine

(a)   the braking torque induced

(b)   the maximum possible acceleration of the flywheel

6.18   The maximum braking force which a car can safely apply is related to the friction between the wheels and the road.

(a)   If a car of mass 850 kg must be able to stop from a speed of 40 km/h in a maximum distance of 100 m, determine the minimum coefficient of friction between the road and the tyres. You may assume the applied force is constant and that forces are evenly distributed over all four wheels.

(b)   If the wheels themselves are 350 mm in diameter, determine the braking torque required.

(c)   The brakes themselves are disc brakes and act on a disc of nominal diameter 225 mm. They utilise friction pads whose coefficient of friction is 0.5. Determine the compressive force on the pad–disc interface in order to achieve the required braking torque.

(d)   Determine the work done by the braking force. How is this energy dissipated in a real system?

6.19   A mechanical hoist operates by an electric motor rotating a 0.5 m diameter drum. A cable is attached to the drum. The other end of the cable is attached to a platform on which the load is placed. When the drum rotates one way, the cable 'wraps around' the drum and the load is lifted. When it rotates the other way, the load is lowered. The maximum load the hoist can carry is 1 tonne. The platform itself has a mass of 100 kg. The platform is further supported by runners similar to these in Problem 6.7, which induce a constant resistance to motion equivalent to a force of 10 kN. If the platform is to accelerate to its operating speed of 1 m/s in 1.5 s, determine

(a)   the acceleration of the platform

(b)   the tension in the cable required to induce the acceleration determined in (a)

(c)   the torque produced by the electric motor

(d)   the nominal power of the electric motor

Chapter 7

Thermofluid applications

The interaction between mechanical work and heat was introduced in Chapter 5. Now we examine fundamental principles of thermodynamics and fluids. This chapter will look at the states of matter, analysis of perfect gases and basic fluid dynamics.

7.1 Basic units

Thermodynamics and fluid dynamics use specific units, and this section introduces them.

7.1.1 Density, volume and mass

For any body, its dimensions are related to its mass by

$$m = \rho V \tag{7.1}$$

where

ρ = density of the material (kg/m^3)

V = volume of the object (m^3)

The density of a material can be considered as the amount of matter which is contained within a given space. Hence light materials such as air and helium have low densities; heavy materials such as steel or lead have high densities. Some typical density values are given in Table 7.1.

Table 7.1 Typical density values for some common materials

Material	Density (kg/m³)	Material	Density (kg/m³)
Cork	86	Aluminium	2 700
Oil	890	Mild steel	7 850
Ice	928	Lead	11 340
Water	1 000	Mercury	13 600
Dry sand	1 750	Gold	19 300

Example 7.1   *A 50 mm square-section steel bar is 2 m long. Determine its volume and its mass.*

SOLUTION
The volume of the bar is given by its cross-sectional area multiplied by its length:

$$V = Al$$
$$= (0.05 \times 0.05) \times 2$$

hence

$$V = 0.005 \, \text{m}^3$$

The mass of the bar is given by

$$m = \rho V$$
$$= 7850 \times 0.005$$

hence

$$m = 39.25 \, \text{kg}$$

7.1.2   Fluid flow

For any fluid, of density ρ, flowing with speed v in a channel or tube of cross-sectional area A (Figure 7.1), the fluid flow may be described by its velocity v. The characteristics of flow may also be described using volumetric flow rate and mass flow rate.

Volumetric flow rate

$$\dot{V} = Av \qquad\qquad (7.2)$$

Mass flow rate

$$\dot{m} = \rho\dot{V} = \rho Av \qquad\qquad (7.3)$$

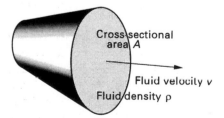

Cross-sectional
area A

Fluid velocity v
Fluid density ρ

Figure 7.1   Fluid flow in a pipe or channel

The units of volumetric flow rate are m^3/s and for mass flow rate kg/s. BS 5775 gives mass flow rate as q_m and volumetric flow rate as q_v. However, most practising engineers recognise the use of $\dot{m}$ and $\dot{V}$, so we shall adopt them in this text.

Example 7.2 *A pipe of diameter 25 mm transports water at a nominal flow velocity of 2.5 m/s. Determine the volumetric flow rate and mass flow rate of the water.*

SOLUTION
The cross-sectional area of the pipe is given by

$$A = \frac{\pi D^2}{4}$$
$$= \frac{\pi (0.025)^2}{4}$$

hence

$$A = 0.49 \times 10^{-3}\, m^2$$

Recalling equations (7.2) and (7.3), volumetric flow rate is given by

$$\dot{V} = A v$$
$$= 0.49 \times 10^{-3} \times 2.5$$

hence

$$\dot{V} = 1.22 \times 10^{-3}\, m^3/s$$

Mass flow rate is given by

$$\dot{m} = \rho \dot{V}$$
$$= 1000 \times 1.22 \times 10^{-3}$$

hence

$$\dot{m} = 1.22\, kg/s$$

7.1.3 Temperature

The *Celsius* scale (commonly called centrigrade) is used for temperature measurment. However, this is a relative scale based on the ice point and the boiling point of water; the interval between is divided into 100 divisions (hence centi). The ice point and the boiling point of water vary considerably with pressure. For example, measurements at sea level do not compare with measurements in a mountainous region. Relative temperature scales use the symbol t and are scalar quantities.

Because the Celsius scale is variable, engineers and scientists use *absolute temperature* as the basis of their measurements. The scale is based on *absolute zero*, the temperature at which *all* motion stops, even the motion of atoms themselves. It has been found that a graph of temperature versus volume for a gas is a straight line, as shown in Figure 7.2. If the straight line is extrapolated to the temperature axis, it crosses at the same point for *all* gases. The temperature at which this occurs is $-273\,°C$ (actually $-273.15\,°C$ but the last two decimal places are normally immaterial for engineering calculations). The absolute temperature scale uses the symbol T and is also a scalar quantity. The unit for absolute temperature is the kelvin (K).[1] The divisions of the scale have been selected to match the divisions of the Celsius scale. Thus $0\,°C$ is equivalent to 273 K and $100\,°C$ to 373 K.

---

[1] After William Thomson, 1st Baron Kelvin of Largs (1824–1907).

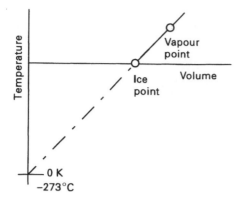

Figure 7.2   Evaluation of absolute zero

Example 7.3   *Convert the following to absolute temperature*

(a) 100 °C
(b) −25 °C
(c) 30 °F

SOLUTION
(a) $T = 100 + 273 = 373\,\text{K}$ $(0\,°\text{C} = +273\,\text{K})$
(b) $T = -25 + 273 = 248\,\text{K}$
(c)   We must convert the now out-of-date Fahrenheit scale to centigrade. To do this we use the conversion

$$°\text{C} = (°\text{F} - 32)/1.8$$

To obtain absolute temperature we add 273 K, hence

$$T = [(°\text{F} - 32)/1.8] + 273$$
$$= [(30 - 32)/1.8] + 273$$
$$= 271.89\,\text{K}$$

7.1.4 Pressure

Consider a column of fluid of height h and cross-sectional area A, as illustrated in Figure 7.3. The body force exerted on the base by the mass of the fluid in the column is given by

$$F = mg$$

where the mass of the fluid in the column is given by

$$m = \rho V = \rho h A$$

Atmospheric pressure

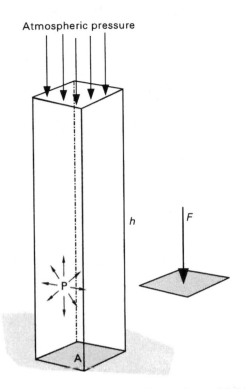

Figure 7.3 Pressure induced by a column of fluid

Hence the body force exerted on the area A is given by

$$F = \rho g h A$$

Hydrostatic pressure p is defined as the force per unit area exerted by a motionless fluid, hence

$$p = F/A$$
$$= \rho g h A / A$$

therefore

$$p = \rho g h \qquad\qquad (7.4)$$

Equation (7.4) describes the hydrostatic pressure generated by a fluid at depth h. In any fluid p is a scalar; furthermore, p acts in all directions (consider the pressure in a rubber balloon). Care must be taken with this term because p can be relative to atmosphere, *gauge pressure*, or relative to zero, *absolute pressure*:

Absolute pressure = gauge pressure + atmospheric pressure $\qquad (7.5)$

Atmospheric pressure, P_o, is the pressure induced by the air surrounding us. Commonly this is about $100\,\text{kN/m}^2$, which is sometimes called 1 bar. Atmospheric pressure is variable and should always be measured as a matter of course.

Example 7.4   *Determine the absolute pressure at the bottom of a 25 m well which is full of water.*

SOLUTION
Recalling equation (7.4) we note that the hydrostatic pressure is given by

$$p = \rho g h$$
$$= 1000 \times 9.81 \times 25$$

hence

$$p = 245.52 \, kN/m^2 \, (or \, kPa)$$

But the surface of the water is open to the atmosphere, so the absolute pressure at the bottom of the well is given by

$$p = \rho g h + p_0$$
$$= 242.25 \times 10^3 + 100 \times 10^3$$

hence

$$p = 342.25 \, kN/m^2$$

Example 7.5   *Some gauges give pressure in terms of height of fluid. Convert the following to absolute pressure:*
   (a) 60 mm of mercury (60 mmHg)
   (b) 250 mm of water

SOLUTION
Recalling the solution to Example 7.4. The absolute pressure in a column of fluid open to the atmosphere is given by

$$p = \rho g h + p_0$$

Hence for part (a)

$$p = (13\,600 \times 9.81 \times 0.06) + 100\,000$$
$$= 108 \, kN/m^2$$

And similarly for part (b)

$$p = (1000 \times 9.81 \times 0.25) + 100\,000$$
$$= 102.45 \, kN/m^2$$

Example 7.6   *The hydraulic braking system of a car relies upon a brake-pedal lever applying a force to a piston of diameter 12.5 mm. The piston in turn pressurises a fluid, which applies force to another piston of diameter 25 mm. This in turn applies a force to the braking pad. Determine the amplification of force by this system.*

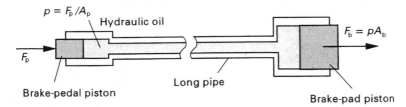

Figure E7.6

SOLUTION

Figure E7.6 illustrates the problem. If we assume the fluid in the pipe is incompressible, the pressure within the fluid will be constant throughout. The pressure generated at the brake-pedal piston is given by

$$p = F_p/A_p$$
$$= F_p/(\pi \times 0.006\,25^2)$$

hence

$$p = F_p/122.7 \times 10^{-6}$$

This is the same as the pressure at the brake-pad piston, where $p = F_b/A_b$, so

$$p = F_b/(\pi \times 0.0125^2)$$
$$= F_b/490.9 \times 10^{-6}$$

hence

$$p = F_p/122.7 \times 10^{-6} = F_b/490.9 \times 10^{-6}$$

which, when rearranged, yields the amplification factor F_b/F_p

$$F_b/F_p = 490.9 \times 10^{-6}/122.7 \times 10^{-6}$$

or

$$F_b/F_p = 4$$

Hence this braking system effectively increases the force at the brakes by 4; the system has a mechanical advantage of 4.

Pressure has the potential to do work, so the stored energy of a fluid at a pressure p is called *pressure energy* or *flow work*. Consider a pressure p acting on one side of a plate and a force F acting on the opposite side, as depicted in Figure 7.4. If the pressure moves the plate by a distance s, the work done on the force by the fluid is

$$W = Fs$$

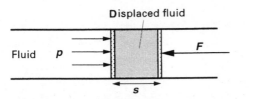

Figure 7.4   Plate acted upon by pressure p and force F

The work done by the fluid is also *Fs*, but $F = pA$, hence

$$W = pAs$$

Since *As* is the volume of the displaced fluid, the work done by the fluid is given by

$$W = pV$$

The work done by or to a fluid is also represented by the area under a graph of pressure versus volume (often called a *pV diagram*), as illustrated by Figure 7.5.

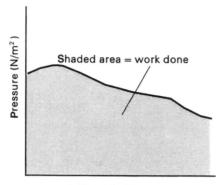

Figure 7.5   Typical *pV* diagram

In the situation where *p* is not constant, work is defined as

$$W = \int p dV$$

which is the general expression for work done due to pressure.

7.1.5   Heat

Chapter 5 considered the relationship between heat and the temperature change of a body; see equation (5.8). For a given material, the change in temperature can also produce a change in volume or a change in pressure. It is therefore common to consider processes where one variable is kept constant, i.e. a constant-pressure process or a constant-volume process. Thus equation (5.8) is rewritten as

$$Q = mc_p \Delta T \tag{7.6}$$

where c_p is the specific heat capacity for a constant-pressure process. Alternatively

$$Q = mc_v \Delta T \tag{7.7}$$

where c_v is the specific heat capacity for a constant-volume process. In practice the values of c_p and c_v vary with temperature, so they are determined by experiment. Their values depend on the material in question, but at $T = 0\,\mathrm{K}$, $c_p = c_v = 0$ for all materials – one reason why $0\,\mathrm{K}$ is so difficult to achieve! The units of specific heat capacity are J/kg K.

This text considers only systems which are heated (or cooled) at a constant atmospheric pressure. Table 7.2 gives typical values of c_p for some common materials, in ascending order.

Table 7.2 Typical values of specific heat capacity

Material	c_p (J/kg K)
Lead	129
Pure copper	385
Mild steel	440
Stainless steels	470
Pure aluminium	903
Softwoods	1 630
Paraffin/kerosene	2 000
Water	4 180

Example 7.7 *Determine the heat required to change the temperature of a bath containing 90 litres of water from 20°C to 50°C.*

SOLUTION
Assuming the water is heated at constant pressure, we can use equation (7.6):

$$Q = mc_p \Delta T$$

We know that the mass of the water is ρV, from Section 7.1. Also we recall that 1 litre is $1000 \times 10^{-6}\,m^3$, hence

$$m = 1000 \times (90 \times 1000 \times 10^{-6})$$

thus

$$Q = (1000 \times 90 \times 1000 \times 10^{-6}) \times 4180 \times (50 - 20)$$

thus

$$Q = 11.3\,MJ$$

Unfortunately, this simple question has a major error hiding within it; we have neglected the fact that we must also heat the bath itself!

7.2 The three states of matter

Any material can be attributed to three specific states; *solid, liquid* and *gas*. The states of matter depend on the boundary conditions at that instant. To illustrate this concept, Figure 7.6 depicts a state diagram for an arbitrary material; the vertical axis is pressure and the horizontal axis is temperature. This diagram is in fact three-dimensional – the third axis is volume – but for simplicity the third axis has been removed.

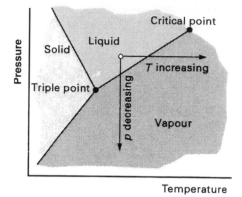

Figure 7.6   State diagram for a material

Note that the term *fluid* is not indicated. This is because both liquids and vapours (or gases) are called fluids.

The lines indicate where a material changes state. When it crosses the liquid–vapour line the material changes from a liquid to a vapour, or vice versa. The point where it crosses is often called the *boiling point* or *vapour point*. When it crosses the liquid–solid line the material changes from a liquid to a solid, or vice versa. The point where it crosses is often called the *ice point*. The liquid-to-solid transformation is called melting.

The change of state of a material due to a decrease in pressure is illustrated by the vertical line drawn on Figure 7.6. Temperature is kept constant, so the process is *isothermal*. As can be seen, the material changes state from a liquid to a vapour. This phenomenon is exhibited at high altitudes. Here the atmospheric pressure is so reduced that water boils at a much reduced temperature, compared with that obtained at sea level.

Where a material is subject to an increase in temperature (at constant pressure), exemplified by boiling a kettle, the material changes state from fluid to vapour. This is illustrated by the temperature-increasing line on Figure 7.6.

The *triple point* is very specific. It is a condition where all three states of matter are exhibited at the same time, i.e. the material is solid, liquid and gaseous. For a given material it only occurs at one specific condition defined by a unique pressure, volume and temperature. The *critical point* is where there is no longer any quantifiable or discernible difference between the liquid and gaseous states.

7.3   Perfect and ideal gases

After a series of extensive experiments, Boyle[2] stated:

The pressure of a fixed mass of gas is inversely proportional to its volume if retained at the same temperature.

[2] Robert Boyle, English natural philosopher and chemist (1627–1691).

Boyle's law was extended to model any perfect gas, whose characteristic equation may be written as

$$pV = mRT$$

where R is the *gas constant* for a particular gas[3] and has the units J/kg K. Table 7.3 illustrates gas constants for some common gases.

Table 7.3 Values of gas constant for typical gases

Gas (molar mass)	Gas constant, R (J/kg K)
Carbon dioxide (44)	189
Propane (44)	189
Oxygen (32)	259.8
Carbon monoxide (28)	296.9
Nitrogen (28)	296.9
Methane (16)	519.6
Hydrogen (2)	4 157

In practice, gases do not fully obey Boyle's law and the characteristic equation. However, for gases of low density (such as oxygen, helium and nitrogen at pressures below 50 bar and temperatures between 200 and 3000 K) the correlation between theory and real life is close enough to make equation (7.8) useful. Gases which are in this category are called *semi-perfect* or *ideal gases*.

Example 7.8 *A quantity of propane of mass 0.2 kg is to be stored in a rigid 200 litre container. If the nominal storage temperature is 15 °C, determine the storage pressure of the gas.*

SOLUTION
The container is rigid, so we can assume that its dimensions do not change as the gas is pumped into it. Applying equation (7.8), we get

$$pV = mRT$$

or

$$p = \frac{mRT}{V}$$
$$= \frac{0.2 \times 189 \times (15 + 273)}{(200 \times 1000 \times 10^{-6})}$$
$$= 54 \, kN/m^2 \quad \text{or} \quad 0.54 \, bar$$

Note that this is absolute pressure.

---

[3] The gas constant may be determined from $R = \dfrac{R_0}{M}$, where R_0 is the universal gas constant (8314.3 J/kmol K) and M is the molar mass of the gas.

7.4   Buoyancy in fluids

Archimedes[4] discovered the following law of buoyancy which now bears his name:

> *The buoyancy force felt by a body which is either fully or partially immersed in a fluid is equal to the weight of the fluid displaced.*

And *Archimedes' principle* may be written as

$$F_b = g\rho_f V'$$
(7.9)

where F_b is the the buoyancy force, ρ_f is the density of the fluid, and V' is the volume of fluid displaced. The buoyancy force is illustrated in Figure 7.7.

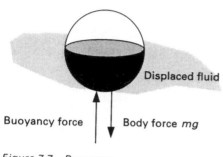

Displaced fluid

Buoyancy force   Body force *mg*

Figure 7.7   Buoyancy

Example 7.9   *A small flat-bottomed dinghy is designed to support four passengers, a total mass of 340 kg. If the surface area of the dinghy, A = 2 m², can be assumed to be constant for all water depths, determine the draught of the vessel.*

SOLUTION
Firstly, the term *draught* means the minimum depth of water required by the vessel in order to float, that is the vertical distance from the keel to the water-line. Figure E7.9(a) illustrates this in more detail.

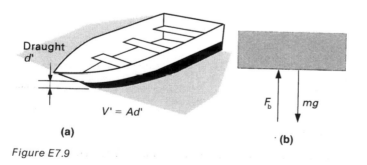

Draught
d'

$V' = Ad'$

F_b mg

(a)

(b)

Figure E7.9

---

[4] Archimedes published his work on buoyancy and the principle of the lever in the third century BC; this is considered to be the first ever scientific publication.

Recalling equation (7.9) and examining the free-body diagram of the vessel (Figure E7.9(b)) reveals that

$$\Sigma F = 0$$

hence

$$F_b - mg = 0$$

or

$$F_b = mg$$

That is, for a body to float, the buoyancy force and the body force must be in equilibrium. When equation (7.9) is substituted for the buoyancy force F_b, we get

$$g\rho_f A d' = mg$$

which can be rearranged to yield an expression for draught d':

$$d' = \frac{mg}{g\rho_f A} = \frac{m}{\rho_f A}$$

The density of water is $1000 \, \text{kg/m}^3$, thus

$$d' = \frac{340}{1000 \times 2}$$

hence

$$d' = 0.17 \, \text{m}$$

Example 7.10 *If the sidewalls of the dinghy are 0.21 m high, determine the maximum load the dinghy can carry. You may assume the unladen mass of the dinghy to be 12 kg.*

SOLUTION
The dinghy will sink when the water level exceeds the height of its sidewalls. Hence the maximum draught of the dinghy is $d' = 0.21 \, \text{m}$. Applying equation (7.9) allows the determination of the maximum buoyancy force:

$$F_b = g\rho_f A d'$$

hence

$$F_b = 9.81 \times 1000 \times 2 \times 0.21$$
$$= 4120 \, \text{N}$$

Thus the maximum load the dinghy can carry is 4120 N or 420 kg. Which, when the mass of the dinghy is taken into account, reveals that the maximum additional load is 408 kg.

7.5 The continuity equation

Consider a fluid flowing through a pipe, as illustrated in Figure 7.8. The principle of *continuity of mass* states that the mass flow rate at point 1 is identical to the mass flow rate at point 2, irrespective of change of section or change of direction:

$$\dot{m}_1 = \dot{m}_2 \qquad\qquad (7.10)$$

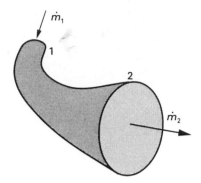

Figure 7.8   Continuity of mass

which may be written as

$$\rho_1 A_1 v_1 = \rho_2 A_2 v_2 \tag{7.11}$$

For an *incompressible fluid* $\rho_1 = \rho_2$. Typical incompressible fluids are liquids such as water and oil; compressible fluids are gases such as air.

Example 7.11   *A junction in a pipe is illustrated in Figure E7.11. Oil enters the pipe at A with a mass flow rate of 15 kg/s. On exit at C the oil has a flow velocity of 12 m/s. Determine the mass flow rate and velocity of flow at exit pipe B. The internal diameters of the pipes at A, B and C are 25 mm, 18 mm and 12 mm respectively.*

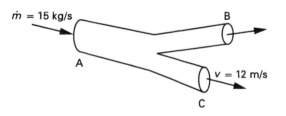

Figure E7.11

SOLUTION

From the principle of continuity of mass, we can write

$$\dot{m}_A = \dot{m}_B + \dot{m}_C$$

which we can rearrange for the mass flow rate at B

$$\dot{m}_B = \dot{m}_A - \dot{m}_C$$

We know the mass flow rate at A, but we only know the flow velocity at C. Recalling equation (7.3) allows us to write

$$\dot{m}_B = \dot{m}_A - \rho A v_C$$

or

$$\dot{m}_B = 15 - \left[1000 \times \left(\frac{\pi \times 0.012^2}{4}\right) \times 12\right]$$

hence

$$\dot{m}_B = 13.643\,\text{kg/s}$$

Rearranging equation (7.3) allows to write an expression for the flow velocity at B:

$$v = \frac{\dot{m}_B}{\rho A}$$

or

$$v = \frac{13.643}{1000 \times (\pi \times 0.018^2/4)}$$

hence

$$v = 53.61\,\text{m/s}$$

7.6 Conservation of momentum

The principle of conservation of momentum is a very powerful tool in fluid dynamics, but fluid dynamicists tend to treat momentum differently to other disciplines because they deal with *fluid flow* instead of a moving solid. For a fluid hitting a curved plate (Figure 7.9) the mass flow rate is constant but the direction of flow changes, hence the fluid velocity changes.

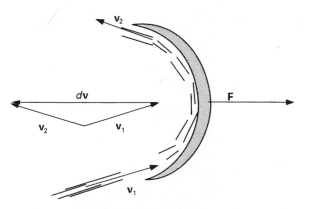

Figure 7.9   Change in direction of flow

As we learnt from Chapters 2 and 4, if velocity changes there must be an acceleration, and if there is an acceleration there must also be a force. The plate has applied a force to the fluid to change its direction. From Newton's third law there must be a reaction to this force. This is the force of the fluid on the plate.

The force acting on the fluid is given by the rate of change of momentum:

$$\mathbf{F} = \frac{(m\mathbf{v})_2 - (m\mathbf{v})_1}{t}$$

but we know that mass flow rate is constant, so

$$\mathbf{F} = \frac{m(\mathbf{v}_2 - \mathbf{v}_1)}{t} = \frac{m}{t}(\mathbf{v}_2 - \mathbf{v}_1)$$

hence

$$\mathbf{F} = \dot{m}(\mathbf{v}_2 - \mathbf{v}_1) \tag{7.12}$$

Equation (7.12) yields the magnitude and direction of the force acting on the fluid. The force acting on the plate has the same magnitude but opposite direction.

Example 7.12   *A boat is powered by a water jet. The water is drawn into the front of the boat and accelerated by a pump which ejects the fluid out of a nozzle, diameter 35 mm, at the rear of the boat. If the water is ejected at a nominal flow velocity of 12 m/s, determine the acceleration of the boat from rest. The mass of the boat is 125 kg.*

SOLUTION

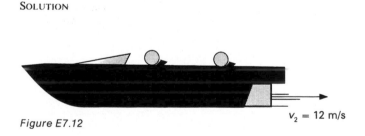

$v_2 = 12$ m/s

Figure E7.12

Figure E7.12 is a schematic of the water jet. The water entering the vessel can be assumed stationary, and the water exiting the vessel has been accelerated to 12 m/s. For this to have occurred, the vessel must accelerate the water, and the water must itself exert an equal and opposite force on the boat. Hence we can write equation (7.12) as

$$\mathbf{F} = \dot{m}(\mathbf{v}_2 - 0)$$

We can determine the mass flow rate from equation (7.3):

$$\dot{m} = \rho A v$$

$$= 1000 \times \left(\frac{\pi \times 0.035^2}{4} \right) \times 12$$

hence

$$\dot{m} = 11.54 \,\text{kg/s}$$

Thus we can determine the force acting on the fluid, which is also the force acting on the boat. This force is

$$F = 11.54 \times (12 - 0)$$

hence

$$F = 138.53\,\text{N}$$

From Newton's second law (applied to dynamics) we know that $\Sigma F = ma$, hence

$$\begin{aligned}
a &= \frac{F}{m} \\
&= \frac{138.53}{125} \\
&= 1.108\,\text{m/s}^2
\end{aligned}$$

Example 7.13 *Figure E7.13(a) illustrates a typical nozzle for a hose-pipe. If the mass flow rate of the water flowing through the pipe is 2 kg/s, determine the force acting on the nozzle. What is this force doing to the nozzle?*

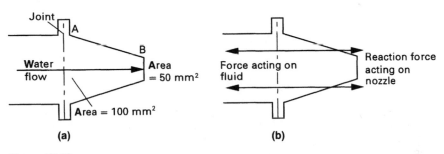

(a)   (b)

Figure E7.13

SOLUTION
At A, the velocity of flow is given by

$$v = \frac{\dot{m}}{\rho A}$$

or

$$v = \frac{2}{1000 \times 100 \times 10^{-6}}$$

hence

$$v_1 = 20\,\text{m/s}$$

At B the velocity of flow is

$$v = \frac{2}{1000 \times 50 \times 10^{-6}}$$

hence

$$v_2 = 40\,\text{m/s}$$

The magnitude of the force acting on the fluid is given by equation (7.12) as

$$F = \dot{m}(v_2 - v_1)$$
$$= 2 \times (40 - 20)$$

thus

$$F = 40 \, \text{N}$$

The direction of the force clearly opposes the flow. Something must have applied a force to the fluid, and that something is the nozzle. Thus the nozzle feels the reaction to the force acting on the fluid, and this reaction is trying to force the nozzle from the hose, as illustrated in Figure E7.13(b). Note that this does not require any knowledge of the shape of the nozzle!

7.7   Conservation of energy and Bernoulli's equation

Consider a fluid flowing through a system as depicted in Figure 7.10. At position 1 the fluid has potential energy mgZ due to its position (note that Z is used to denote height), kinetic energy due to its flow $1/2mv^2$ and pressure energy pV. At position 2 the values of potential, kinetic and pressure energy may have changed but the total must remain constant; we can write this in the form of energy per unit mass (kJ/kg):

$$gZ_1 + \frac{v_1^2}{2} + \frac{p_1 V_1}{m} = gZ_2 + \frac{v_2^2}{2} + \frac{p_2 V_2}{m}$$

which in turn can be rearranged to

$$Z_1 + \frac{v_1^2}{2g} + \frac{p_1}{\rho g} = Z_2 + \frac{v_2^2}{2g} + \frac{p_2}{\rho g}$$

which is also known as *Bernoulli's equation*.[5] Bernoulli's equation is often used to describe the principles of *flight* and *lift*. In this text we shall be examining its implications for fluid flow.

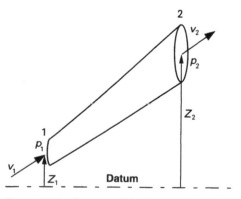

Figure 7.10   Conservation of energy in a fluid

---

[5] After Daniel Bernoulli, Dutch-born Swiss mathematician (1700–1782).

Example 7.14 *A storage tank, open to atmosphere, holds water to a depth of 4 m. Three taps on the side of the tank are situated at the base, 1 m from the base and 2 m from the base respectively. Determine the maximum velocity of flow available from each tap.*

SOLUTION

Let us start with a sketch of the problem (Figure E7.14). We must also allocate known values of p, v and Z at important points (the three taps and the surface of the water).

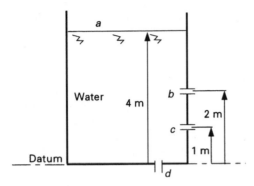

Figure E7.14

At a (the free surface) the pressure is atmospheric pressure, from our datum we have Z +4 m, and the velocity of flow is assumed negligible. Hence

$$p_a = 100\,\text{kN/m}^2$$
$$Z_a = 4\,\text{m}$$
$$v_a = 0\,\text{m/s}$$

Similarly, at b, we also know the pressure is atmospheric because it is exiting into open space. Z is +2 m but we do not know the velocity of flow. Hence

$$p_b = 100\,\text{kN/m}^2$$
$$Z_b = 2\,\text{m}$$
$$v_b = ?\,\text{m/s}$$

But we can apply equation (7.13) to determine v_b:

$$Z_a + \frac{v_a^2}{2g} + \frac{p_a}{\rho g} = Z_b + \frac{v_b^2}{2g} + \frac{p_b}{\rho g}$$

$$4 + \frac{0^2}{2g} + \frac{100 \times 10^3}{\rho g} = 2 + \frac{v_b^2}{2g} + \frac{100 \times 10^3}{\rho g}$$

Note that the free-surface velocity term is zero and the two pressure terms cancel out:

$$4 = 2 + \frac{v_b^2}{2g}$$

or

$$v_b = \sqrt{2g(4-2)}$$
$$= 6.26\,\text{m/s}$$

A close examination of the solution shows that there is a general expression for the velocity of flow from a source open to the atmosphere whose exit (be it a tap, pipe or hole) is at height h below the free surface:

$$v = \sqrt{2gh}$$

For exit point c

$$p_c = 100 \, \text{kN/m}^2$$
$$Z_c = 1 \, \text{m}$$
$$v_c = ? \, \text{m/s}$$

hence the velocity of flow there is given by

$$v_c = \sqrt{2g(4-1)}$$
$$= 7.67 \, \text{m/s}$$

For your own practice show that the velocity of flow at d is $v_d = 8.86 \, \text{m/s}$.

Example 7.15   *Figure E7.15 illustrates a tank of water which is being emptied by a syphon. If the pipe diameter is 25 mm, determine the mass flow rate of the water at exit. Also determine the pressure in the fluid at the apex, indicated at point b. What is the maxium height of point b from the surface of the water so that flow is ensured?*

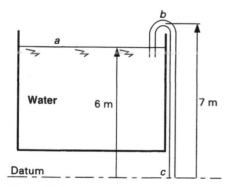

Figure E7.15

SOLUTION

As with Example 7.14, we list the known boundary conditions at points a, b and c:

	p	v	Z
a	100×10^6	0	6
b	?	?	7
c	100×10^6	?	0

Since point *c* has only one unknown, this is clearly the starting point. Furthermore, since the flow is from an exit open to the atmosphere, we can use the general expression determined in Example 7.14 giving

$$v_c = \sqrt{2g(6-0)}$$
$$= 10.85 \, \text{m/s}$$

From continuity of flow, the mass flow rate at the exit *c* must be the same as at point *b*. If we assume that the diameter of the pipe is constant throughout its length, the velocity of flow at *b* must also be the same. Hence

$$v_b = 10.85 \, \text{m/s}$$

By applying equation (7.13) to points *b* and *c* we achieve

$$0 + \frac{10.85^2}{2g} + \frac{100 \times 10^3}{\rho g} = 7 + \frac{10.85^2}{2g} + \frac{p_b}{\rho g}$$

We can rearrange this to obtain an expression for p_b:

$$p_b = \rho g \left(\frac{100 \times 10^3}{\rho g} - 7 \right)$$

Substituting the density of water as $1000 \, \text{kg/m}^3$ and $g = 9.81 \, \text{m/s}^2$, we get

$$p_b = 31.33 \, \text{kN/m}^2$$

This is less than atmospheric pressure, so a pressure gauge would see this as $-68.67 \, \text{kN/m}^2$. The flow will cease when the absolute pressure at *b* becomes zero, hence

$$0 = \rho g \left(\frac{100 \times 10^3}{\rho g} - Z_b \right)$$

which is when the term within the brackets is zero, so

$$Z_b = \frac{100 \times 10^3}{\rho g}$$
$$= 10.19 \, \text{m}$$

Summary

Units

$$\text{Density: } \rho = \frac{m}{V} \, (\text{kg/m}^3)$$

$$\text{Volumetric flow rate: } \dot{V}(q_v) = Av \, (\text{m}^3/\text{s})$$

$$\text{Mass flow rate: } \dot{m} \text{ (or } q_m) = \rho\dot{V} = \rho Av \, (\text{kg/s})$$

$$\text{Pressure: } p = \rho g h \, (\text{N/m}^2)$$

$$\text{Absolute pressure} = \text{gauge pressure} + \text{atmospheric pressure}$$

$$\text{Absolute temperature: } T = {}^\circ\text{C} + 273 \, (\text{K})$$

Gas laws

$$\text{Boyle's law: } pV = \text{constant}$$
$$\text{Characteristic equation: } pV = mRT$$

Heat

$$\text{Constant pressure process: } Q = mc_p\Delta T$$
$$\text{Constant volume process: } Q = mc_v\Delta T$$

Buoyancy

Archimedes' principle states that the buoyancy force exerted by a body when immersed in a fluid is equal to the weight of fluid displaced:

$$F_b = g\rho_f V'$$

Continuity of flow

$$\dot{m}_1 = \dot{m}_2 \quad \text{or} \quad \rho_1 A_1 v_1 = \rho_2 A_2 v_2$$

Conservation of momentum

$$\mathbf{F} = \dot{m}(\mathbf{v}_2 - \mathbf{v}_1)$$

Conservation of energy using Bernoulli's equation

$$Z_1 + \frac{v_1^2}{2g} + \frac{p_1}{\rho g} = Z_2 + \frac{v_2^2}{2g} + \frac{p_2}{\rho g}$$

Problems

Basic principles

7.1   A container stores $13\,\text{m}^3$ of air whose density is $1.22\,\text{kg/m}^3$ at atmospheric pressure. Determine the mass of air stored.

7.2   Determine the volume of container required to store $44.7\,\text{kg}$ of lubricating oil of density $894\,\text{kg/m}^3$.

7.3   Convert the following temperatures to absolute temperature:
   (a) $100\,°\text{C}$
   (b) $0\,°\text{C}$
   (c) $0\,°\text{F}$

7.4   A pipe of diameter $25\,\text{mm}$ conveys a fluid of specific density 0.8 with a velocity of flow $v = 10\,\text{m/s}$. Determine the volumetric flow rate and mass flow rate of the fluid. Note: specific density is density relative to the density of water.

7.5   A $10\,\text{m}$ column of water is open to the atmosphere. Determine
   (a)   the gauge pressure at the base of the column
   (b)   the absolute pressure at the base of the column

7.6 Table P7.6 illustrates the temperaure pressure data for a particular gas (volume is kept constant). Draw a graph of temperature versus pressure and obtain the absolute temperature when the pressure is zero. What is this temperature known as?

Table P7.6

Temperature (°C)	Gauge pressure (MN/m^2)
0	0.98
25	1.06
50	1.16
75	1.25
100	1.35

7.7 A piston of diameter 40 mm is open to the atmosphere on one side and there is an absolute pressure of $80 kN/m^2$ on the other side. Determine the magnitude and direction of the resultant force acting on the piston.

7.8 A 2 litre domestic kettle is used to heat water from a nominal tap water temperature of 10 °C to boiling point in 2.5 minutes. Determine the heat and power required to carry out this task.

7.9 Convert the following gauge pressures to absolute pressure:
 (a) $450 kN/m^2$
 (b) 80 mm mercury
 (c) 15 m water

7.10 A container of volume $0.05 m^3$ contains nitrogen stored at a gauge pressure of 5 bar and at a nominal 20 °C.
 (a) Determine the mass of nitrogen stored.
 (b) If the storage temperature is reduced by 10 °C, determine which variable alters and determine its new value.

7.11 Which of the following are true statements related to the triple point?
 (a) All states of matter occur at the same time.
 (b) The triple point can exist at any pressure.
 (c) The triple point occurs at unique conditions of p, T and V for a given material.
 (d) Water does not have a triple point.

7.12 Which of the following statements describes a gas which obeys Boyle's law?
 (a) The pressure of a gas is directly proportional to its volume.
 (b) If pressure increases volume decreases.
 (c) If two different gases are stored in similar vessels, their mass being identical, then their pressures are identical.
 (d) For a perfect gas $pV = $ constant.

7.13 Which of the following statements accurately describes the principles of the characteristic equation for a gas?
 (a) For any gas the value of mRT is always the same.
 (b) For a given gas, if temperature changes then pressure must also change.
 (c) For a constant-volume process, if temperature increases then pressure must increase.
 (d) For any gas, $pV = $ constant.
 (e) The characteristic equation is valid for any gas whatever its density.

Buoyancy

7.14   A steel buoy of mass 125 kg is constructed such that its total volume is $0.5\,m^3$. Will the vessel float in water?

7.15   A child's balloon, the mass of rubber being 2.5 g, contains helium ($\rho = 0.179\,kg/m^3$), its total volume being 3 litres. Determine the 'pull' on the string, due to buoyancy, which the child feels ($\rho_{air} = 1.225\,kg/m^3$).

7.16   A small boat of mass 1.25 tonnes is to be used inshore. Determine the minimum amount of water it must displace in order to get afloat.

Continuity of flow

7.17   A joint in a piping system splits one 25 mm diameter pipe into two pipes of diameter 12.5 mm. The pipe transmits water, and the velocity of flow into the main pipe is 13 m/s. If the mass flow rate out of one of the 12.5 mm pipes is 2 kg/s, determine
   (a)   the mass flow rate of water entering the joint
   (b)   the mass flow rate passing through the remaining 12.5 mm pipe
   (c)   the volumetric flow rates in all the pipes

7.18   In a housing estate one 100 mm diameter pipe supplies all the domestic water to eight houses. The peak demand for water in each of the houses is 120 litres per minute. Determine
   (a)   the minimum volumetric flow rate of water which the main pipe must carry
   (b)   the peak mass flow rate of water to each house

7.19   A joint in a piping system connects three separate pipes. The two inlet pipes have mass flow rates 9.7 kg/s and 7.9 kg/s respectively. A third pipe transmits the water with a flow velocity of 12 m/s. In an effort to smooth the mixing at the joint, the flow velocities for all pipes, inlet and outlet, are made the same. Determine
   (a)   the diameters of the two inlet pipes
   (b)   the mass flow rate of water out of the joint
   (c)   the diameter of the outlet pipe

Conservation of momentum

7.20   The paddles on a water-wheel are perfectly flat. A jet of water from a 12.5 mm diameter nozzle hits the paddle, and most of the water is turned through 90°, as shown in Figure P7.20. If the velocity of flow from the nozzle is 25 m/s, determine the maximum force the jet exerts on the paddle, assuming the paddle is stationary.

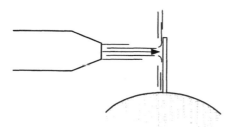

Figure P7.20

7.21   What factors would you have to consider for the solution of Problem 7.20 if the paddles were moving?

7.22 A horizontal jet of water hits a wall, inclined at +45° to the ground, and is turned up the wall. If the water jet has a velocity of 25 m/s and a mass flow rate of 30 kg/s, determine the magnitude and direction of the force acting on the wall.

7.23 By how much is the force increased on the paddle of Problem 7.20 if the paddles are redesigned such that the water is turned through an angle of 180°?

Conservation of energy

7.24 An A4 sheet of paper has been folded into a top-hat profile, as depicted in Figure P7.24, and placed on a smooth tabletop. Describe what happens if someone blows through the channel. (You may wish to try this for yourself. Another good example is to make a cone, put a ping-pong ball into it then try to blow it out!)

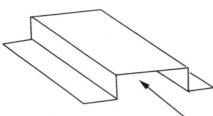

Figure P7.24

7.25 A bathroom water-tap is supplied from the mains at a gauge pressure of 10 m of water (at ground level). Determine the maximum velocity of flow at the tap if it is situated 2.1 m above ground level.

7.26 In an effort to increase the velocity of flow of water in a pipe, the pipe diameter was reduced from 50 mm to 40 mm. The supply conditions to the 50 mm section of the pipe are gauge pressure 2 bar and mass flow rate 25 kg/s. If the pipe is level, determine the pressure in this 40 mm section.

7.27 A tank stores fuel oil ($\rho = 860 \text{ kg/m}^3$). The depth of the tank is 2 m and a small tap is placed at its base. The tank stands on supports of height 0.5 m. If the diameter of the tap outlet is 12.5 mm, determine the mass flow rate of the fuel oil when the tap is fully open.

7.28 In order to empty it at a faster rate, the tank of Problem 7.27 is raised on blocks to a new height of 2 m. Determine the new mass flow rate from the tap.

7.29 Determine the mass flow rate of fuel oil from a syphon placed in the tank of Problem 7.27 if the free end of the syphon is at ground level and the pipe has a 25 mm diameter.

Chapter 8

Structural analysis

The term *structure* is defined as follows:

> **Structure: an arrangement of parts or elements to form another object whose purpose is to carry or transmit a load.**

So in structural analysis we model and analyse a whole structure by looking at its components. This chapter will apply to structures of the basic principles we have developed so far. In particular, it will introduce beam analysis and basic framework analysis.

8.1   Beam analysis

A beam is any single long, slender structural component whose purpose is to support an applied load. It must transmit this load sucessfully from the point of application to a point of support. It is subject to forces which are *normal* to its axial length, as illustrated in Figure 8.1. Commonly these forces are coplanar. Structural components which are loaded *axially* are examined further in Section 8.2. Application of the coplanar forces puts the beam into *bending*. But for this to arise there must exist one of the following conditions:

(a)   If only forces are applied, there must at least three.
(b)   If only moments are applied, there must be at least two.
(c)   At least two forces and one moment must be applied.

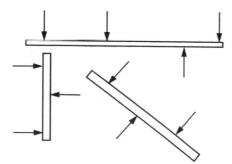

Figure 8.1   Three typical beams

In this section we shall be modelling beams analytically to obtain known and unknown forces using static equilibrium. Also we shall obtain shear force and bending moment relationships for the beam. But first we must introduce the fundamental components of beam analysis. We must examine the various supports, loading methods and beam types.

8.1.1 Applied load definitions

Beams are commonly subject to four types of load: point load, uniformly distributed load, non-uniformly distributed load and applied moment. Their definitions are as follows.

Point loads
A point load is a force which acts on a single infinitesimally small point. An example of a point load is illustrated in Figure 8.2. This type of load is exemplified by objects supported on hangers. A point load is drawn on the beam's free-body diagram as a single arrow at the point of application. Note that the free-body of the beam is drawn as a thick line; the reason for this will become apparent when we consider distributed loads.

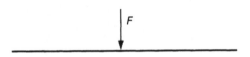

Figure 8.2   Representation of a point load

Distributed loads
Distributed loads model the forces induced by the beam's mass or any load which acts over a large proportion of the beam. There are two forms of distributed load, *uniformly distributed (UDL)* and *non-uniformly distributed.*

On the free-body diagram the uniformly distributed load is drawn as a box (Figure 8.3), its length representing the section of the beam over which it acts. Often a uniformly distributed load is allocated the symbol w and is given as load per unit length, N/m.

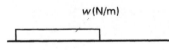

Figure 8.3   Representation of an applied UDL

The non-uniformly distributed load is drawn as an irregular box (Figure 8.4) indicating the section of the beam over which it acts; the actual loading is presented as an equation where the load is a function of distance, $F = f(x)$. The study of non-uniformly distributed loads is beyond the scope of this text, but it is important that you know the difference.

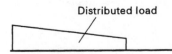

Figure 8.4   Representation of a non-uniformly distributed load

Applied moment

A moment or a torque may be applied directly to a beam. This is represented by an arrow which indicates the magnitude and sense of the applied moment (Figure 8.5). In beams it is common to assume that a clockwise moment is positive. This convention has been selected so that the effect of applied moments to a beam's bending moment is consistent with other load types.

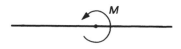

Figure 8.5   Representation of an applied moment

8.1.2   Support definitions

There are numerous methods of supporting a beam, and hence numerous models. Here we shall meet the three most common: *knife-edge* or simple support, *pin* support and *built-in* support. The supports are examined on the basis of the movement which they allow. The forces induced by the support limit the motion of the beam, hence it is beneficial to categorise a support by the motion that *does not occur*.

Knife-edge or Simple support

Figure 8.6 illustrates a typical knife-edge supporting a beam. The beam cannot move downwards because the support limits this motion, but it can move up, side to side, or spin about the point of contact.

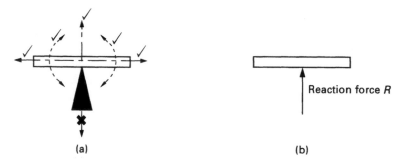

(a)   (b)

Figure 8.6   Knife-edge or simple support: (a) schematic and (b) support reaction

Because the support limits motion downwards there must be a force to create static equilibrium; this is called the support *reaction force*, and is commonly given the symbol R. The direction of the force is normal to the beam. To restrict the motion vertically upwards as well as downwards, two knife-edges can be mounted directly opposed, as illustrated in Figure 8.7.

In reality only one vertical reaction force exists at a time, but it can act in either direction to resist the applied load.

Note that friction between the support and the beam has been treated as negligible.

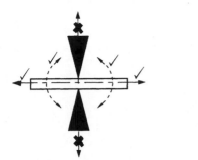

Figure 8.7 Double knife-edge

Pin-joint

Figure 8.8 illustrates a typical pin-joint. Commonly the joint consists of a round shaft attached to the beam running in frictionless bearings. Hence the beam cannot move vertically up or vertically down. Nor can it move from side to side. It can, however, spin about the point of rotation.

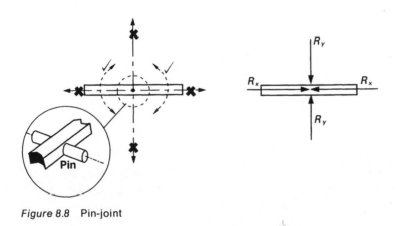

Figure 8.8 Pin-joint

Hence the support can apply both an axial reaction force and a normal reaction force to the beam. The reaction force R_x can exist in either direction, but only one at a time. This is also the case for R_y.

Often a pin-joint may be simplified so it can be modelled in a similar manner to the simple support, that is with a normal reaction force (R_y) only.

Built-in support

As the name suggests, the beam is considered as being rigidly fixed to a structure (Figure 8.9). In this case the beam cannot move in any direction. The support induces reaction forces in both axial and normal directions. It also induces a reaction moment to restrict the rotation of the beam.

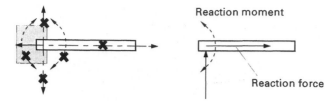

Figure 8.9   Model of a rigid support

In all of the previous examples the horizontal component R_x has been included in the discussion. In practice, beams are subjected to loads which can be at any orientation, so the R_x component must exist. In this text we shall only consider beams with normal loads, hence the axial component, R_x, may be ignored.

8.1.3   *Common types of beam*

Balance or scales
This type of beam is characterised by a single support, normally central, which acts as a fulcrum. The combination of two or more forces either side of the fulcrum creates a state of balance, as depicted in Figure 8.10. A small version of this beam may be found in kitchen scales.

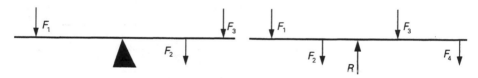

Figure 8.10   Examples of 'balanced' beams

Simply supported beam
This type of beam relies on two supports creating a state of static equilibrium, as illustrated in Figure 8.11. The supports can be knife-edges or pin-joints.

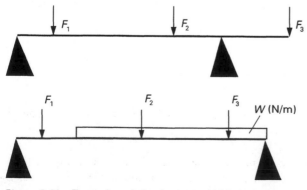

Figure 8.11   Examples of simply supported beams

Cantilever beam

A beam which is built-in at one end and free at the other is called a *cantilever*, as illustrated in Figure 8.12. Cantilevers are very common and are found supporting hanging baskets and shelves all over the world.

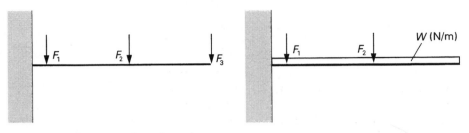

Figure 8.12 Examples of cantilever beams

Encastre beam

This type of beam is built-in at both ends, as shown in Figure 8.13. It is commonly found in civil engineering structures as a joist or lintel.

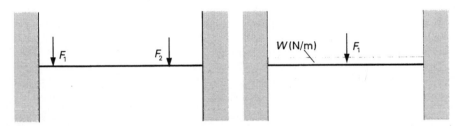

Figure 8.13 Examples of encastre beams

8.1.4 Determining reaction forces and moments

In Section 8.1.2 we saw that, in order to fulfil its function, each support must apply reaction forces and moments to the beam. In this section we shall be examining the method of obtaining the reactions.

First we recall that the beam must be static, in which case Newton's second law states

$$\Sigma \mathbf{F} = 0$$

and

$$\Sigma \mathbf{M} = 0$$

The forces and moments are often along the same axis, so vector notation may be ignored. If the beam is statically determinate, the application of Newton's second law is sufficient to obtain all reaction forces.

If the beam is statically indeterminate, which we will not be studying, other boundary conditions must be defined. These are normally based around known deflections at a support.

Example 8.1   *Determine the reaction forces at the left-hand and right-hand supports for the beam illustrated in Figure E8.1(a).*

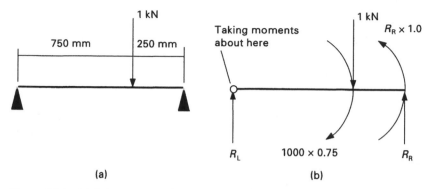

Figure E8.1   (a) Schematic and (b) free-body diagram

SOLUTION

Firstly we recognise that the supports are knife-edge supports. Hence they exert a vertical reaction force only. We assume that they 'point upward', this is our positive direction. Our calculations will tell us whether we were right.

Taking moments about the left-hand support and applying Newton's second law leads to an equation which has removed one of the unknowns ($R_L \times 0 = 0!$). Figure E8.1(b) illustrates a free-body diagram of the beam with the induced moments included. Taking clockwise moments as positive we obtain

$$\Sigma M = 0$$

$$(1000 \times 0.75) - (R_R \times 1.0) = 0$$

or

$$(1000 \times 0.75) = (R_R \times 1.0)$$

Note that this is the same as stating:

All clockwise moments equal all anticlockwise moments.

We can now determine the reaction at the right-hand side:

$$R_R = \frac{750}{1.0}$$
$$= 750 \, \text{N}$$

To obtain the reaction at the left-hand support, we apply Newton's second law to forces:

$$\Sigma F = 0$$

$$-1000 + R_R + R_L = 0$$

hence the reaction at the left-hand support must be

$$R_L = 250 \, \text{N}$$

A close inspection of Example 8.1 illustrates an analytical short cut available to the engineer. If a beam (Figure 8.14), which has two simple supports, is subject to an applied force F, the contribution of that force to the left-hand reaction is

$$R_L = \frac{Fb}{l} \qquad\qquad (8.1)$$

and the contribution to the right-hand reaction is

$$R_R = \frac{Fa}{l} \qquad\qquad (8.2)$$

where

a = distance from the applied load to the left-hand support (positive to the left)
b = distance from the applied load to the right-hand support (positive to the right)
l = distance between supports

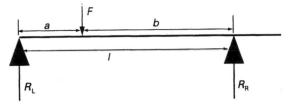

Figure 8.14 Contribution of F to R_L and R_R

This may be repeated for as many loads as the beam carries. The resulting separate values of R_L may be added together to yield the total reaction force, and similarly for R_R. This is called the *principle of superposition*. It may even be used for beams which have overhung loads.

Example 8.2 *Figure E8.2(a) illustrates a simply supported beam carrying three point loads and a UDL. Determine the reactions* R_L *and* R_R.

SOLUTION
Recalling equations (8.1) and (8.2), we can apply them to the three point loads and the UDL in turn.
For the 2 kN force

$$R_L = \frac{2 \times 3.5}{4}$$
$$= 1.75\,kN$$
$$R_R = \frac{2 \times 0.5}{4}$$
$$= 0.25\,kN$$

In a similar manner, for the 3 kN force, we find that

$$R_L = 1.125\,kN$$
$$R_R = 1.875\,kN$$

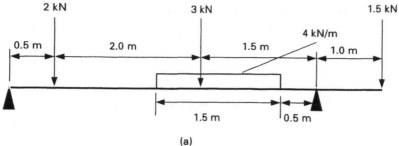

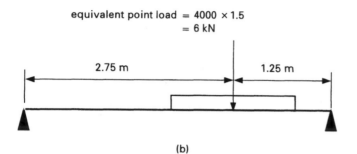

equivalent point load = 4000 × 1.5
= 6 kN

Figure E8.2 (a) Schematic of beam and (b) equivalent point load for the UDL

For the overhung load, 1.5 kN, we can also use equations (8.1) and (8.2):

$$R_L = \frac{1.5 \times (-1.0)}{4}$$
$$= -0.375 \, \text{kN}$$

Note that the distance to the right-hand support, b, is negative.

$$R_R = \frac{1.5 \times 5}{4}$$
$$= 1.875 \, \text{kN}$$

To analyse the UDL, we first determine an *equivalent point load*. The equivalent point load for a UDL is a point load equal in magnitude to the UDL but which acts at the UDL's midpoint. For the UDL in this example

$$F = 4000 \times 1.5 = 6 \, \text{kN}$$

and the equivalent point load acts at a distance of 2.75 m from the left-hand end, as illustrated in Figure E8.2(b). Thus the contribution of the UDL to the support reactions is

$$R_L = \frac{6 \times 1.25}{4}$$
$$= 1.875 \, \text{kN}$$
$$R_R = \frac{6 \times 2.75}{4}$$
$$= 4.125 \, \text{kN}$$

The total reaction at each support is therefore

$$R_1 = 1.75 + 1.125 - 0.375 + 1.875$$
$$= 4.375 \, \text{kN}$$
$$R_R = 0.25 + 1.875 + 1.875 + 4.125$$
$$= 8.125 \, \text{kN}$$

Example 8.3 *Figure E8.3 illustrates a cantilever beam subject to two point loads and a UDL. Determine the reaction at the wall.*

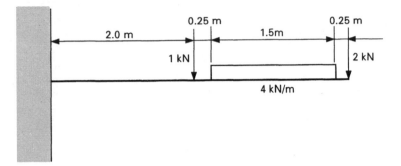

Figure E8.3

SOLUTION
Firstly we recognise that the support at the wall must be a built-in support. Therefore there must be both a reaction force (R) and a reaction moment (Mo) at this point. Cantilevers are, in fact, easier to deal with than other types of beam. We still apply Newton's second law to the beam, but it is much easier to solve!

There is only one support, so it must react to all moments and forces. Therefore

$$Mo = -\Sigma M$$

Treating clockwise moments as positive, we have

$$Mo = [(1000 \times 2) + (2000 \times 4) + (4000 \times 1.5 \times 3)]$$
$$= -28 \, \text{kN m} \quad \text{(or 28 kN m anticlockwise)}$$

and

$$R = -\Sigma F$$
$$= -[-1000 - 2000 - (4000 \times 1.5)]$$
$$= 9 \, \text{kN} \quad \text{(or 9 kN upwards)}$$

8.1.5 Shear force and bending moment
The structural performance of a beam is characterised in a number of ways, however, a starting point in all beam analysis is the derivation of shear force and bending moment relationships. The derivation can be either graphical or algebraic. In this text the graphical solution will be utilised, purely because it will allow you to picture beams in your mind. This will help you when you begin to study the analytical methods.

A beam's function is to transfer applied loads, over a distance, from the point of application to a point of support. The combination of all applied loads often leads to one section of a beam carrying a greater load than another section. It is important to determine the location and magnitude of maximum loads in order to 'design out' the possibility of beam failure.

Consider a beam which we imagine to be cut into two pieces. If we examine the forces acting on the mating surfaces then the left-hand section applies a force *F* downwards on the right-hand section. The system was in static equilibrium before we made the cut, so the right-hand section must also induce an equal and opposite force on the left-hand section. A very thin slice of beam juxtaposed in this gap (Figure 8.15) will therefore have a combination of equal and opposite forces on its faces.

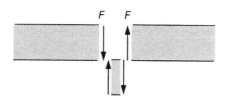

Figure 8.15   Shear force in a beam

This combination of an equal and opposite force in a beam is called *shear force*. Any object which is subject to a force system, as that exerted on the central section, is said to be subject to a *shearing force* or shear. The forces combine to attempt to shear the material rather like a pair of shears (or scissors) cutting paper, dress material, hair and grass! The sign convention defines clockwise shear as positive, anticlockwise as negative (Figure 8.16).

Figure 8.16   Sign convention for shear

Because they are separated by distance, the applied loads also induce localised moments on a slice of the beam. Figure 8.17 illustrates a section of a beam subject to the applied moments caused by *all* loads acting on the beam.

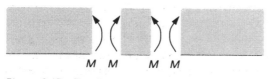

Figure 8.17   Bending moments in a beam

The beam is in static equilibrium, therefore equal and opposite moments must exist on either side of the slice. This combination of two moments causes the beam to bend, as shown in Figure 8.18, hence they are called *bending moments*.

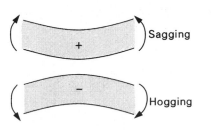

Sagging

Hogging

Figure 8.18 Convention of a bending moment

The sign convention is as follows. If the bending moment causes the beam to *sag*, it is positive. If the bending moment causes the beam to *hog*, it is negative. An easier way to remember this rule is that a sagging beam resembles a smile, which is a positive attitude.

8.1.6 Shear force relationships.

Shear force is a measure of the forces acting on a left-hand section of a beam compared to those on a right-hand section. It is a simple numerical sum which can be carried out using a *shear force diagram*. Consider the simply supported beam illustrated in Figure 8.19. If we begin at the left-hand end, the shear force diagram is established by graphical addition. At each point load, the value of the load is added to the shear force to the left of the load. Each UDL can be considered as a series of point loads. The UDL is continuous, so the 'steps' are smoothed to a straight line. The diagram is completed by noting that between the applied loads there is no change in shear force. Simply follow the arrows!

Before drawing the diagram, an important boundary condition should be remembered. At the *free ends*, the beam is not connected to anything. Because it is not connected, nothing can transmit a force to it, nor can it transmit a force to anything. Hence the shear force must be zero at free ends. If there are applied loads or supports at each end, this fact still applies because in reality we can never apply any load at the very edge of a beam.

Once complete, you should realise that the shear force diagram is a graph. At any point on the beam, the shear force can be obtained by reading the corresponding value on the graph. It is important to highlight the positive and negative regions. Figure 8.19 shows that the portion of the beam which extends from the left-hand end to the 200 N load is the portion that carries the greatest shear force, +240 N.

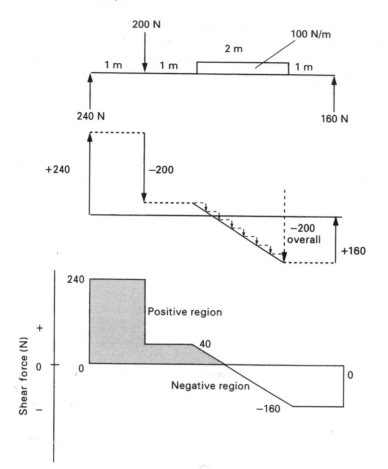

Figure 8.19   Shear force diagram for a simple beam

8.1.7 Bending moment relationships

The bending moment diagram is directly linked to the shear force diagram:

For any section of a beam, the change in bending moment over that section is equal to the area of the shear force diagram carried by that section.

Consider the beam illustrated in Figures 8.19 and 8.20. We can easily identify five areas on the shear force diagram (Figure 8.20): A, B, C, D and E. We now calculate each area individually:

$$A = 240 \times 1 = 240$$

$$B = 40 \times 1 = 40$$

$$E = -160 \times 1 = -160$$

Areas C and D are slightly more difficult. First we have to determine the point where the shear force line crosses the zero axis. We know that the UDL is 100 N/m and that the shear force at the start of the UDL is 40 N. Hence the distance from the start of the UDL to where the shear force is zero may be determined from

$$x = F/w$$

where F is the value of the shear force at the beginning of the UDL and w is the value of the UDL in N/m. Thus the point where the shear force is zero is

$$x = \frac{40}{100} = 0.4\,\text{m}$$

which is 2.4 m from the left-hand end of the beam. We can now determine areas C and D:

$$C = 40 \times 0.4 \times 0.5 = 8$$
$$D = -160 \times 1.6 \times 0.5 = -128$$

The bending moment diagram may now be drawn. This is carried out by adding the values of the areas in sequence as the beam is traversed from left to right. As with the shear force diagram, at the free ends of the beam there cannot be any bending moments.

As a guide, vertical dotted lines have been drawn on Figure 8.20, and they delimit the boundaries of the five areas we chose earlier. By simple addition we can find the value of the bending moment at each of these vertical lines. But we do not know what happens in between.

If the region in question connects two point loads, the bending moment relationship is a straight line. If a UDL is present, the bending moment relationship is a curve. In this text the exact nature of the curve[1] is not important because we know the value of the maximum and minimum values.

In this example the point where the shear force diagram crosses the zero shear force axis is the point of maximum bending moment. This is often the case and it is a fact worth remembering. In this beam the maximum bending moment is 288 N m and is 2.4 m from the left-hand end. Also the bending moment is positive over the entire length of the beam, which suggests that it sags.

The art of drawing shear force and bending moment diagrams is to be able to represent the relationships as a sketch rather than as a technical drawing. Quite often a quick sketch of the system can help in the engineering design process. In later studies you will meet methods of describing these quantities as equations.

---

[1] If you wish to be exact, the curve can be plotted knowing that the change in bending moment is given by $0.5wa^2$, where w is the value of the UDL and a is the distance from the left-hand end of the UDL.

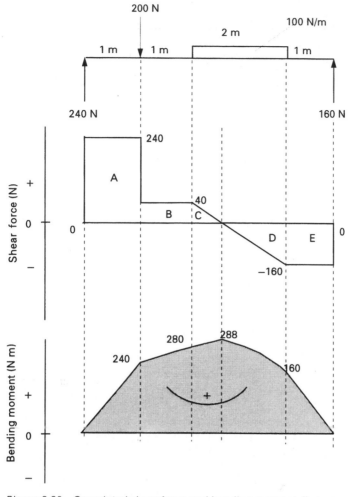

Figure 8.20 Completed shear force and bending moment diagrams

---

Example 8.4 *Draw shear force and bending moment diagrams for the beam illustrated in Example 8.3.*

SOLUTION

First of all we recognise the problem as a cantilever beam, and this simplifies the analysis. For the shear force diagram, the starting value is the reaction force at the support. For the bending moment, the starting value is the reaction moment at the support. At the free end, both shear force and bending moment must be zero. It couldn't be easier!

The shear force diagram can be drawn using the 'follow the arrows' principle described earlier. The change in bending moment for each section is established for areas A, B, C and D:

$$A = 9 \times 2 = 18 \, \text{kN m}$$
$$B = 8 \times 0.25 = 2 \, \text{kN m}$$
$$C = (0.5 \times 6 \times 1.5) + (2 \times 1.5) = 7.5 \, \text{kN m}$$
$$D = 2 \times 0.25 = 0.5 \, \text{kN m}$$

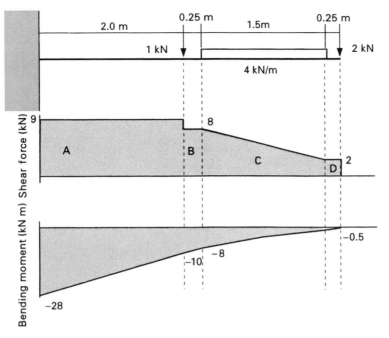

Figure E8.4

Note that all the changes are positive, so to finish at zero the bending moment diagram must start at a negative number; this is the reaction moment at the support. The diagrams are sketched in Figure E8.4.

8.2 Framework analysis

Another component commonly used in engineering is the pin-jointed member. Pin-jointed members can be found in many products, ranging from lawnmowers through to racing-car suspensions. Their widespread use is often attributed to their analytical simplicity.

8.2.1 Definition of a pin-jointed member

A pin-jointed member is a component which is connected to the outside world via pin-joints. Commonly they are long bars which can be round in section and have pin-, rose- or ball-joints at either end, as shown in Figure 8.21. The joints allow the member to spin freely about its fixing but keep it located axially. Thus only axial forces can exist, and moments cannot be transmitted. The axial forces transmitted can either stretch or squash the member; they are known as *tensile* and *compressive* forces. The sign convention is positive for tensile forces, because they increase the member's length. Compressive forces are therefore negative. Members that are in tension are called *ties*, members in compression are called *struts*.

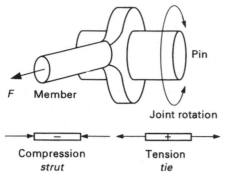

Figure 8.21   Typical pin-jointed member

8.2.2   Definition of a framework.

A framework is a construction of a number of pin-jointed struts and ties used to transmit forces from their points of application to a point of support. The use of many struts and ties can often produce a rigid system which uses little material, as shown in Figure 8.22. Since the pin-joints cannot support applied moments – they are free to spin about a fixing – it is very easy to design a framework which is not in static equilibrium!

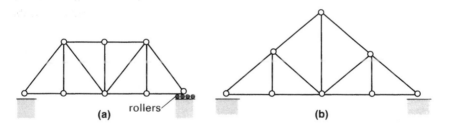

Figure 8.22   Two typical frameworks: (a) Warren bridge and (b) Howe roof-truss

A framework is often constructed from a number of members forming triangles. In this case the *stability* of the system can be estimated by examining the number of joints (*j*) and the number of members (*m*).

Type I, $m + 3 < 2j$
The framework will collapse and requires more members for stability.

Type II, $m + 3 = 2j$
The framework is statically determinate and should therefore be stable.

Type III, $m + 3 > 2j$
The framework is overdesigned because it contains *redundant* components and is statically indeterminate.

8.2.3 Modelling frameworks

In this section we shall be examining frameworks of type II only. Before we begin the analysis we must examine how the framework is supported.

Frameworks have to be mounted to a surrounding structure. Since all the joints are pin-joints, the mounting points can also be pin-joints. Then reaction forces are all that exist at the framework supports, but their direction may not always be known. Another common feature is that one mounting point is connected to ground via rollers, as illustrated in Figure 8.22(a). These allow the end of the framework to move, usually to accommodate expansion on hot days or shrinkage on cold days (see Chapter 9). Any mounting point which uses rollers can only exhibit a *normal reaction force*.

We can now begin to analyse a framework in order to determine forces in frame members. This can be carried out in one of two ways: the method of joints and the method of sections.

Method of joints

For a joint to be in static equilibrium, all forces must sum to zero. Thus the whole framework is modelled by considering the free-body diagram of each joint, as depicted in Figure 8.23. Note (from Newton's third law) that the joint is subjected to the reaction to the state of force in each member. If a member is in compression the force is 'pushing' on the joint; if a member is in tension the force is 'pulling' on the joint. It is common to assume that all unknown members are in tension (ties) at the beginning of the analysis; the true situation emerges during the calculations.

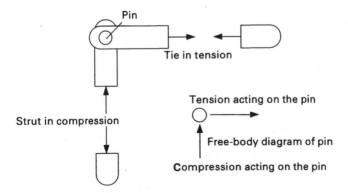

Figure 8.23 Tensile and compressive forces acting on a pin-joint

Example E8.5 *Figure E8.5(a) illustrates a cantilever structure constructed from a simple framework. Each section is made from an equilateral triangle of sides 1, 1 and $\sqrt{2}$. Using the method of joints, determine the magnitude and type of the force in AB and the force in AG. From your calculations show that member BG is under zero load.*

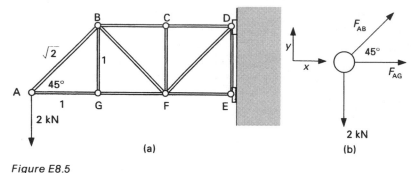

(a)   (b)

Figure E8.5

SOLUTION

Figure E8.5(b) is a free-body diagram of the pin-joint at A, assuming all unknown forces in members to be tensile. We can determine the unknown forces by applying Newton's second law of motion to horizontal and vertical components of force acting on the joint. But first we must check that the framework is of type II:

Number of members, $m = 11$
Number of joints, $j = 7$
$$(11 + 3) = (2 \times 7)$$

so the framework is of type II. We can now analyse the joint in question by examining the horizontal and vertical components of the forces, as denoted by our x–y axes:

Vertically

$$\Sigma F_y = 0$$
$$F_{AB} \sin 45^\circ - 2000 = 0$$
$$F_{AB} = 2.829\,\text{kN}$$

Hence member AB is in tension.

Horizontally

$$\Sigma F_x = 0$$
$$F_{AG} + F_{AB} \cos 45^\circ = 0$$
$$F_{AG} = -2\,\text{kN}$$

Hence member AG is in compression.

Examining joint G shows that, since AG is horizontal, there is no vertical force acting at G, so member BG must be acting under zero load.

Method of sections

For a particular section, not only the forces sum to zero but also the moments. It can be a lengthy process to analyse a whole structure using the method of joints, but the *method of sections* may speed things up. The framework is arbitrarily split so that a 'line of cut' splits up to three unknown members, as illustrated in Figure 8.24. The section of the beam which remains, often to the left of the cut, is treated as a *rigid body*. Newton's second law is then applied to determine any unknown forces.

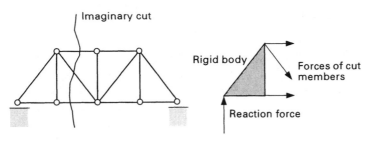

Figure 8.24 The method of sections

Example 8.6 *For the framework described in Example 8.4, use the method of sections to determine forces in members CD, FD and FE.*

SOLUTION
Figure E8.6(a) illustrates the 'cut' required to separate the three members in question. The free-body diagram for the left-hand portion is shown in Figure E8.6(b). In these types of problem it is beneficial to start by taking moments about points which help to remove unknowns.

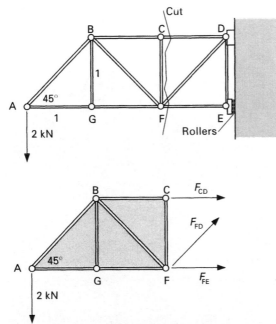

Figure E8.6 (a) Schematic and (b) free-body diagram for left-hand portion

To start with, if we take moments about joint F (within the rigid body), the contribution from members FE and FD can be neglected since they act through F! We apply Newton's second law while taking moments.

Taking moments about F

$$\Sigma M = 0$$
$$(F_{CD} \times 1) - (2000 \times 2) = 0$$
$$F_{CD} = 4\,\text{kN}$$

Taking moments about G (which is not affected by FE but is affected by the vertical component of FD)

$$\Sigma M = 0$$
$$(-2000 \times 1) - (F_{FD} \sin 45^{\circ} \times 1) + (F_{CD} \times 1) = 0$$

hence

$$F_{FD} = -2.828\,\text{kN (compression)}$$

Taking moments about C (which removes CD but includes the horizontal component from FD)

$$\Sigma M = 0$$
$$(-2000 \times 2) - (F_{FD} \cos 45^{\circ} \times 1) - (F_{FE} \times 1) = 0$$

hence

$$F_{FE} = -2\,\text{kN (compression)}$$

Note how the method can be used to obtain forces at this point without knowing the forces in BF, CF or GF!

Summary

Beams
Beam analysis often begins with the drawing of shear force and bending moment diagrams.

Types of support examined
Simple support/knife-edge, vertical reaction force
Pin-joint, vertical reaction force (horizontal may exist as well)
Built-in, vertical and horizontal reaction forces and reaction moment

Types of load examined
Point load
Distributed loads, uniform and non-uniform
Applied moment

Types of beam examined
Simply supported
Cantilever
Encastre

Problems

Beam analysis

8.1   Figure P8.1 illustrates five beams. Determine the reactions at the supports for each beam.

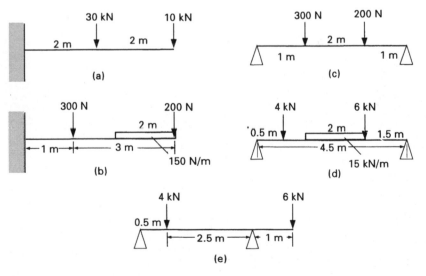

Figure P8.1

8.2   Figure P8.1 illustrates five beams. Draw shear force and bending moment diagrams for each beam. Determine the magnitude and position of the maximum bending moment in each case.

Framework analysis

8.3   Figure P8.3 illustrates a framework. Show that this framework is of type II. Also determine the reaction forces at A and E.

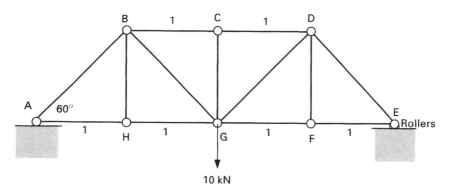

Figure P8.3

8.4   Using the method of joints, determine the forces in members AB, AH and BH for the framework illustrated in Figure P8.3

8.5   Using any method you desire, determine the remaining member forces for the framework illustrated in Figure P8.3.

8.6   Figure P8.6 illustrates a cantilever framework; confirm that it is of type II.

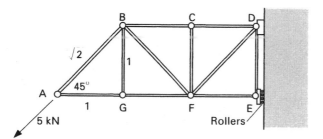

Figure P8.6

8.7   Using the method of joints, determine the forces in members AB, AG and GB.

8.8   Using the method of sections, determine the forces in members BC, BF and GF.

8.9   Using any method you desire, determine the forces in members CD, FD, FE and DE. Hence determine the reactions at D and E.

Behaviour of materials

In previous chapters we saw how to model engineering components and how to use these models to determine the forces acting on them. These models are established at the design stage of a component and used time after time. We need to be sure that a component will not fail in use. One of the most common modes of failure is that the component is not strong enough to withstand the forces applied. The complete opposite is that a component can be made to be too strong – a waste of precious resources. In this chapter we shall be examining the role the material takes in the engineering model.

9.1   Direct stress and strain

When a component is subject to an applied load, as illustrated in Figure 9.1, the force is assumed to be carried evenly by the component. It is said that the force is *uniformly distributed* over the whole cross-section. The ratio of force to cross-sectional area is called *direct stress* and is given the symbol σ and has the units N/m^2. Direct stress is a vector quantity, but in preliminary studies only unidirectional stress is considered, hence it is often used in its scalar form

$$\sigma = \frac{F}{A} \tag{9.1}$$

Cross-sectional area

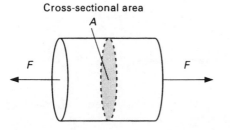

Figure 9.1   Component subject to loads normal to the cross-section

If the applied force is tensile, as illustrated in Figure 9.1, the direct stress is positive. If the force is compressive – a squashing force – the direct stress is negative. Sometimes engineers reduce the name of this quantity to just stress, but there are many forms of stress so it is better to use the full name.

A material will fail when the value of applied direct stress exceeds the maximum value the material can withstand. The strength of a material is often defined by the amount of direct stress it can support.

Example 9.1   *A tie-rod of diameter 40 mm is subjected to a tensile load of 70 kN. Determine the direct stress in the rod.*

SOLUTION
Recalling equation (9.1)

$$\sigma = \frac{F}{A}$$
$$= \frac{70 \times 10^3}{\pi(0.02)^2}$$
$$= 55.7 \, \text{MN/m}^2$$

Note that this answer is positive because the applied load is tensile.

In addition to generating stress within the material, the applied load will also deform the component, as illustrated in Figure 9.2. In the case of a tensile force the component will extend, in the case of compression the component will shrink. The ratio of the change in dimension of the component (in the direction of the applied force) to the original dimension is called *direct strain* and is given the symbol ε. This quantity is dimensionless:

$$\varepsilon = \frac{\Delta l}{l} \tag{9.2}$$

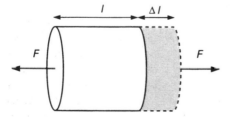

Figure 9.2   Direct strain resulting from an applied tensile load

Quite often students think that direct strain is defined by length only, but remember we live in a 3D world, so an object's length, breadth, width or diameter can change due to an applied force.

Example 9.2   *The tie-rod in Example 9.1 is 1.450 m in length. Under load the tie-rod extends to a length of 1.451 m, determine the direct strain in the tie-rod.*

SOLUTION
The change in dimension of the tie-rod is its change in length, which is

$$\Delta l = 1.451 - 1.450$$
$$= 0.001 \, \text{m}$$

Recalling equation (9.2) yields

$$\varepsilon = \frac{\Delta l}{l}$$
$$= \frac{0.001}{1.450}$$
$$= 689.6 \times 10^{-6}$$

9.2   Thermal strain

The strain attained within a component due to an applied load can also arise from a change in temperature. Imagine the embarrassment caused when, having assembled a complex piece of machinery, it fails during the summer because all of its components have changed size. An increase in the operating temperature of a component often results in an increase in its size.

The expansion of a component may be modelled using

$$\varepsilon = \alpha \Delta T \qquad (9.3)$$

where α is the thermal coefficient of expansion which is described as strain per unit temperature rise. Because strain is dimensionless it is ignored, which makes the unit K^{-1}. Throughout this text we shall use ε/K to reinforce the meaning of α. ΔT is the temperature change which can be in either °C or K. Note that equation (9.3) relates to all three axes (x, y and z) for isotropic[1] materials.

Although this may seem to be a problem, the effects of thermal strain can often be used to advantage. Consider assembling two components which are designed to be pressed together during fitting, as illustrated in Figure 9.3. To reduce the amount of force required for this process, the outer component is often heated by placing it in an oven, hot water or hot oil. Due to thermal strain, the dimensions of this component increase. The diameter of the hole increases and fitting the components together is made easier. Sometimes the same effect is achieved by cooling the inner component using ice, freezers or freezing sprays. After some time, normal steady-state temperatures exist and a tight fit is achieved.

---

[1] Isotropic means that the material has the same properties in all directions. An anisotropic material has different properties in different directions.

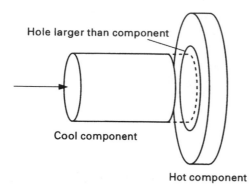

Hole larger than component

Cool component

Hot component

Figure 9.3   Fitting components using thermal strain

Example 9.3   *A component was measured at a temperature of 15°C and a hole was found to have a diameter of 45.01 mm. The component was measured again at an elevated temperature of 150°C, when the hole was found to have a diameter of 45.10 mm. Determine the thermal coefficient of expansion for the material.*

SOLUTION

The dimension that changes is the hole diameter, hence the thermal strain is given by

$$\varepsilon = \frac{\Delta d}{d}$$
$$= \frac{(45.1 - 45.01)}{45.01}$$
$$= 0.002$$

Equation (9.3) provides the solution to this problem:

$$\varepsilon = \alpha \Delta T$$

which when rearranged for α becomes

$$\alpha = \frac{\varepsilon}{\Delta T}$$
$$= \frac{0.002}{(150 - 15)}$$
$$= 14.8 \times 10^{-6} \varepsilon/K$$

Note that when dealing with temperature change use of either °C or K is acceptable since they are the same value.

Example 9.4   *By how much does the length of a 2m steel bar change when its ambient temperature changes from 23°C to −10°C. You may assume α for steel is 12×10^{-6} ε/K.*

SOLUTION

Recalling equation (9.3)

$$\varepsilon = \alpha \Delta T$$
$$= 12 \times 10^{-6} \times (-10 - 23)$$
$$= -0.396 \times 10^{-3}$$

And recalling equation (9.2)

$$\varepsilon = \frac{\Delta l}{l}$$

or

$$\Delta l = \varepsilon l$$
$$= -0.396 \times 10^{-3} \times 2$$
$$= -0.792 \times 10^{-3}\,\text{m}\ (-0.792\,\text{mm})$$

This may not seem much, but consider the implications of the change in length of a railway track!

9.3 Shear stress and strain

In Chapter 8 we met shear force in beams. It is quite common for components to be subject to shear from a number of sources, and this can often produce failure due to shear. The shear force per unit area within a material, where the force acts parallel to the cross-section, as illustrated in Figure 9.4(a), is called *shear stress*. Shear stress is given the symbol τ and has the units N/m^2.

$$\tau = \frac{F}{A} \tag{9.4}$$

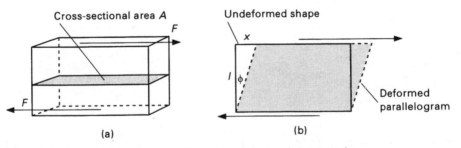

(a) (b)

Figure 9.4 Applied shear force causing shear stress and shear strain

Example 9.5
The tie-rod described in Example 9.1 is connected to the supporting structure by a pin-joint, as illustrated in Figure E9.5(a). The pin is made from 20 mm diameter mild steel rod. Determine the shear stress in the pin.

SOLUTION
A free-body diagram of the pin (Figure E9.5(b)) illustrates that the shear force acting on the pin is one-half of the applied load. This is further illustrated in the shear force diagram

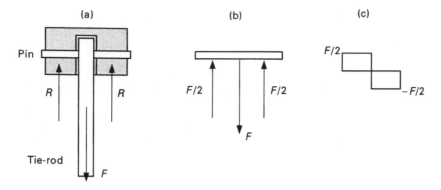

Figure E9.5   (a) Mounting block, (b) free-body diagram and (c) shear force diagram

(Figure E9.5(c)). A component which is loaded this way is often called a *double-shear* component. Recalling equation (9.4)

$$\tau = \frac{F}{A}$$
$$= \frac{35 \times 10^3}{\pi(0.01)^2}$$
$$= 111.4\,\mathrm{MN/m^2}$$

Example 9.6   *The towing-hitch for a tow-bar is illustrated in Figure E9.6. Determine the shear stress in the support bar if the cross-sectional area of the bar is 400 mm² and the maximum applied load F is 10 kN.*

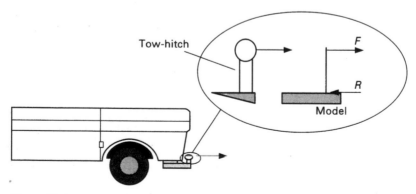

Figure E9.6

SOLUTION
The fixing point can be recognised as a cantilver beam subject to one end-load. From Chapter 8, we know that the shear force is constant and equal to the applied force. Hence the shear stress is given by

$$\tau = \frac{F}{A}$$

$$= \frac{10 \times 10^3}{400 \times 10^{-6}}$$

$$= 25\,\text{MN/m}^2$$

Shear has the effect of distorting a component such that an original rectangular shape, which can be imagined as being etched onto the surface of the component, becomes a parallelogram, as illustrated in Figure 9.4(b). The *shear strain*, γ, is defined by the ratio of the deformation of the parallelogram to the separation of the surfaces x/l:

$$\gamma = \frac{x}{l} \tag{9.5}$$

which, since x/l is always small, is equal to the angular deformation of the parallelogram ϕ.

9.4   Elastic–plastic material properties

Figure 9.5 illustrates a typical materials testing machine. The object of this machine is to apply a tensile load to a specimen of material, whose dimensions are strictly governed by British Standards. These machines are very powerful and they are capable of applying tremendous forces to the specimens. The machine illustrated in Figure 9.5 is able to apply tensile loads which are equivalent to the combined weight of about 150 family cars!

A major benefit of using these testing machines is that we are able to measure both the applied force to the specimen and the extension of the specimen over a given *gauge length* (as shown in Figure 9.6). This enables the materials test engineer to plot graphs of stress versus strain for the material being tested, often called *stress–strain curves*. The specimen has to conform to a testing standard. In Europe the standard which governs the specimen's dimensions is EN 10002-1; for example, the gauge length, L_o of the specimen is given as

$$L_o = 5.65\sqrt{S_o}$$

where S_o is the original cross-sectional area of the specimen.

Note the shape of the specimen; the large ends are usually required so that the testing machine has a suitable amount of area to grip on. The area of concern is often central and, more important, it is distant from the effects of the gripping mechanism. The central portion is often reduced so that the testing machine can break the specimen. The actual extension of the specimen is measured by attaching a measuring device to the specimen which acts over the gauge length.

Figure 9.7 illustrates a typical stress–strain curve for a metal specimen. The vertical axis of the graph is direct stress, and as applied force increases we move up this axis. The horizontal axis is direct strain, and as the specimen extends we move to the right along this axis. The curve depicts the direct strain which exists in the specimen for any given value of direct stress. The abrupt end of the graph is where the specimen finally gave up the fight and failed, perhaps with a bang!

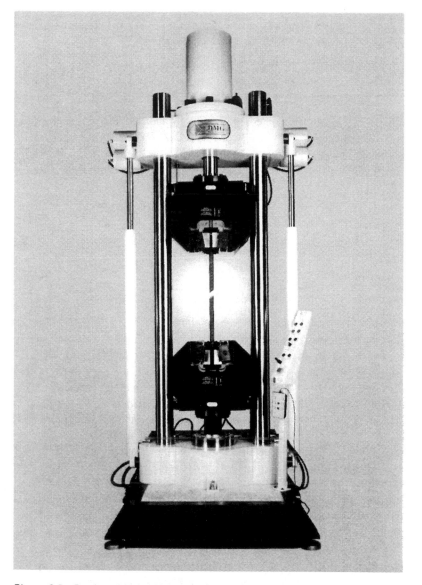

Figure 9.5   Denison Mayes Model 7170: a typical materials testing machine of capacity 1500 kN (Courtesy Denison Mayes Group)

The curve illustrates two significant regions of interest. The first region is the portion of the graph where the relationship between direct stress and strain appears to be linear. This region is called the *elastic region*. Here the material is said to be obeying *Hooke's law*:[2]

For elastic materials the extension of a component is proportional to the force applied.

[2] After Robert Hooke (1635–1703), British philosopher. He was appointed as curator of experiments for the Royal Society in 1662.

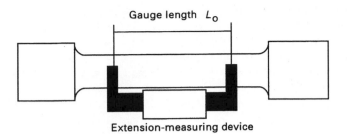

Figure 9.6 A typical tensile test specimen

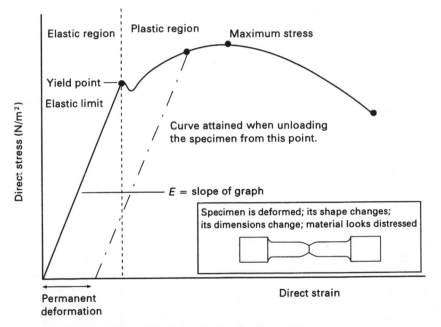

Figure 9.7 Typical stress–strain curve for a ductile material

In our case this is translated to *stress is proportional to strain*. Furthermore, if a material is stretched, extended or expanded within this region it will always return to its original shape once the applied load has been removed.

The gradient of the graph of direct stress versus strain is a constant for a particular material. But it *is* different for different materials. This proportionality is important to engineers because they try to design components to operate in this region. Thus the gradient has been given its own name, *modulus of elasticity*, and the symbol E:

$$E = \frac{\text{direct stress}}{\text{direct strain}} = \frac{\sigma}{\varepsilon} \qquad (9.6)$$

Modulus of elasticity has the units N/m^2, but because materials tend to have moduli of the order of 10^9 the units are often given as GN/m^2. Historically, this quantity is also known as *Young's modulus*.

The second region of interest is called the plastic region. In this region the relationship between direct stress and strain is unclear – it is certainly not linear! The point where the linear relationship between stress and strain breaks down is called the *elastic limit*. Here *permanent*, or *plastic*, deformation begins to occur within the material. The elastic limit is closely followed by *yield*, which occurs at the *yield stress* (σ_y); this is the maximum stress a material can withstand before plastic deformation occurs. In some materials, in particular low-carbon steels, there is marked yield, as illustrated in Figure 9.7. In other materials the yield is not so apparent and may be completely undetectable! Later we shall see methods for estimating a value of yield stress.

For engineers, the yield stress of a material is a very important piece of information. If a component were loaded such that this stress was reached then most engineers would consider their design, or choice of materials, to have failed. It is quite common, in fact, to apply safety factors to a design such that not even one-half σ_y is achieved, let alone the full value!

A material loaded such that it operates in the plastic region will experience *permanent* deformation. The rather bizarre phenomenon that occurs, once the load has been removed, is that the material attempts to return to its orginal shape by obeying Hooke's law. However, the permanent deformation that occurred always remains.

The stress–strain curve will always reach a maximum level of stress, called the *ultimate stress* or *tensile strength* of the material. The final stress which occurs at failure is called the *breaking stress*. Because there is a comparitively large plastic region this material is called *ductile*. Ductility is a relative quantity – one measure is the percentage strain (strain × 100) achieved before failure – but it is an important factor in material choice. The failure of a ductile material is often characterised by local plastic deformation. In particular, components acting under tension often show signs of *necking* as illustrated by the inset to Figure 9.7.

Although most engineering materials have stress–strain curves which are similar to Figure 9.7, there are some cases which are worth noting. Figures 9.8 and 9.9 illustrate two further cases, a material where the yield is undectable and a *brittle* material.

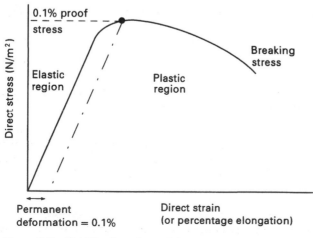

Figure 9.8   Identification of proof stress

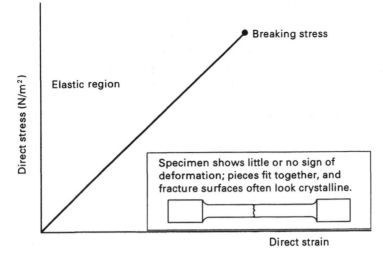

Figure 9.9 Typical stress–strain curve for a brittle material

Figure 9.8 illustrates the stress–strain curve for a particular material where yield is undetectable. There is a clear elastic region and a clear plastic region, but the transition point is unclear. As engineers, we need to assess a material's suitablity for a task, and one quantifiable measure is its yield point. In these situations engineers often refer to the material's *proof stress*. The proof stress of a material is defined as the level of stress required to achieve a fixed amount of permanent plastic deformation (often 0.1% or 0.2%). The proof stress is determined by drawing a line parallel to the elastic portion of the curve. The line starts at the value of permanent plastic deformation and finishes at the actual stress–strain curve. The level of direct stress where this line crosses the stress–strain curve indicates the proof stress of the material.

Figure 9.9 illustrates the relationship between direct stress and strain for a brittle material. Here the plastic region is very small, if it exists at all. The failure of the material is often characterised by a loud snap. On close examination of the broken pieces, the fractured surfaces appear crystalline and often fit back together perfectly. Examples of brittle materials are chalk, tool steels, concrete and ceramics. Brittle materials are often used in compression rather than tension because their tensile properties are so poor in comparison to their compressive properties. It is a bizarre fact that most undergraduates assume that all materials have the same properties in compression as in tension, yet they see chalk and concrete every day!

Example 9.7 *Figure E9.7(a) illustrates a stress–strain graph for a material which was conducted using a tensile test. Determine the modulus of elasticity and the 0.1% proof stress for the material.*

SOLUTION
The definitions for modulus of elasticity E and 0.1% proof stress provide the solution to this problem.

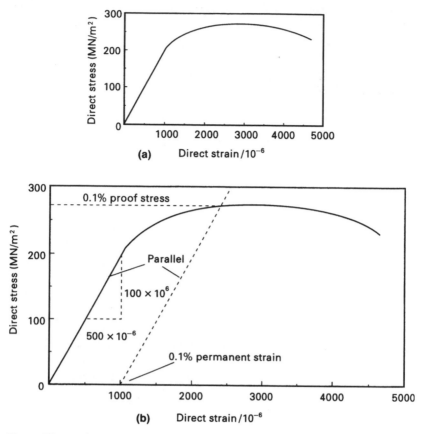

Figure E9.7

To determine E we determine the gradient of the line in the elastic region.

To determine the 0.1% proof stress we draw a line parallel to the elastic portion of the curve beginning at 0.1% strain; 0.1% strain is equivalent to

$$0.1/100 = 0.001 \text{ strain (or } 1000 \times 10^{-6})$$

This is illustrated in Figure E9.7(b).

The gradient of the elastic portion is

$$E = \text{slope}$$
$$= \frac{100 \times 10^{6}}{500 \times 10^{-6}}$$
$$= 200 \text{ GN/m}^{2}$$

and the 0.1% proof stress is given by

$$\sigma_{0.1\%} = \text{stress where the 0.1% line intersects the stress–strain curve}$$
$$\sigma_{0.1\%} = 270 \text{ MN/m}^{2}$$

Example 9.8 *Figure E9.8(a) illustrates several tensile test stress–strain relationships. From the graphs presented determine the following:*

(a) *the material with the greatest modulus of elasticity*
(b) *the material with the smallest 0.1% proof stress*
(c) *the most ductile material*
(d) *the most brittle material*

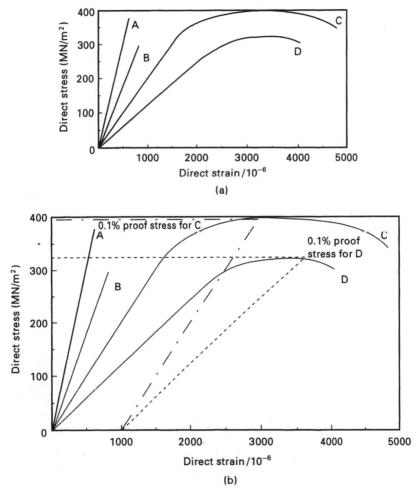

Figure E9.8

SOLUTION

(a) The material with the greatest modulus of elasticity E is the material whose gradient is the greatest within the elastic region. Figure E9.8(a) clearly shows that this is material A.

(b) Figure E9.8(b) illustrates the material properties with the 0.1% permanent strain line added. From the graph it can be seen that the 0.1% proof stresses are as follows:

Material C: $\sigma_{0.1\%} = 395\,\text{MN/m}^2$

Material D: $\sigma_{0.1\%} = 325\,\text{MN/m}^2$

Hence material D has the smallest value of 0.1% proof stress.

(c)   The most ductile material is difficult to ascertain, except subjectively. Firstly we must consider the amount of elongation at failure and the amount of stress which the material withstands. From Figure E9.8(a) we can see that material C has the greater elongation, but material D has the lower ultimate stress. What we can say is that materials C and D are more ductile than A and B; futhermore, material C can be deformed to a greater level than material D, but this will take a greater level of applied stress than for similar deformations of material D.

(d)   Materials A and B are clearly the most brittle. B is more brittle than A because it has a lower stress at failure.

Similar graphs to those presented in Figures 9.7, 9.8 and 9.9 may be achieved for materials subject to shear (i.e. bending of beams or torsion of shafts) by plotting shear stress against shear strain, as in Figure 9.10. The regions of elasticity and plasticity still exist but the relationship in the elastic region is different because the mode of loading is different.

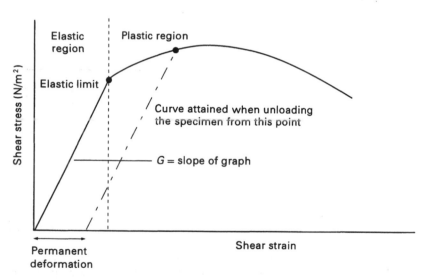

Figure 9.10   Typical shear stress–strain curve for a ductile material

For shear, the gradient of the line in the elastic region is called the *modulus of rigidity* and is given the symbol G, its units are N/m^2.

$$G = \frac{\text{shear stress}}{\text{shear strain}} = \frac{\tau}{\gamma} \tag{9.7}$$

As a rule of thumb, engineers often assume that the yield stress in shear for a material is approximately one-half that of the yield due to direct stress. And for metals, the modulus of rigidity G can be assumed to be 3/8 of the modulus of elasticity E.

To summarise, Table 9.1 lists some commonly used engineering materials with their approximate material properties. When doing any form of design calculation, Table 9.1 can be used as a guide but you should always refer to British Standards, ISO standards or an equivalent for a material's actual properties.

Table 9.1 Some typical material properties

Material	Modulus of elasticity, $E\ (GN/m^2)$	Modulus of rigidity, $G\ (GN/m^2)$	Yield stress, σ_y (MN/m^2)
Mild Steel	200	75	220
Low-alloy steel	209	78	500–1 980
Aluminium	70	26	40
Copper	124	47	60
Titanium	116	44	180–1 320

9.5 Application to engineering problems

Quite often the relationship between stress, strain and a material's properties is used in engineering design. The following examples demonstrate how engineers approach design calculations based around these factors.

Example 9.9 *A tie-rod in a car's suspension is to be constructed from a grade of steel which has a 0.1% proof stress of 250 MN/m². The tie-rod is to be constructed as a solid, round bar and is 350 mm long. If the tie-rod is subjected to a maximum axial force of 10 kN, determine*

(a) the minimum diameter for the tie-rod
(b) the extension of the tie-rod under load (assume E = 209 GN/m²)
(c) the minimum diameter of the tie-rod if a safety factor of 2.5 is applied to the proof stress

SOLUTION
First let us sketch the problem as a free-body diagram (Figure E9.9) of the tie-rod; this illustrates that it is a simple tensile stress problem.

From Newton's laws of motion we know that tensile forces exist at both ends of the rod. Don't let this fool you into a common trap which students make, this *does not* mean that the tie-rod is subjected to a tensile load of 2*F*!

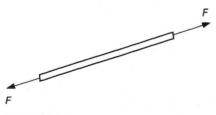

F

F

Figure E9.9

(a)   Recalling equation (9.1)

$$\sigma = \frac{F}{A}$$

which when rearranged for A becomes

$$A = \frac{F}{\sigma}$$

For a round bar, the cross-sectional area A, is given by $A = \pi r^2$. This may be substituted to yield an expression for r:

$$\pi r^2 = \frac{F}{\sigma}$$

$$r = \sqrt{\frac{F}{\pi \sigma}}$$

hence

$$= \sqrt{\frac{10\,000}{\pi \times 250 \times 10^6}}$$
$$= 3.57 \times 10^{-3}\,\text{m (or 3.57 mm)}$$

(b)   Equation (9.6) can be rearranged to yield an expression for ε:

$$E = \frac{\sigma}{\varepsilon}$$

or

$$\varepsilon = \frac{\sigma}{E}$$

hence

$$\varepsilon = \frac{250 \times 10^6}{209 \times 10^9}$$
$$= 1.196 \times 10^{-3}$$

Equation (9.2) reveals the link between extension and strain:

$$\varepsilon = \frac{\Delta l}{l}$$

or

$$\Delta l = \varepsilon l$$
$$= 1.96 \times 10^{-3} \times 0.35$$
$$= 0.419 \times 10^{-3}\,\text{m (or 0.419 mm)}$$

(c)   To apply a safety factor to a yield stress or proof stress, simply divide the corresponding value by the safety factor:

$$\sigma_{SF} = \frac{\sigma_{0.1\%}}{SF}$$

where σ_{SF} is the stress value used in subsequent calculations and SF is the safety factor. Hence in this case

$$\sigma_{SF} = \frac{250 \times 10^6}{2.5}$$
$$= 100 \times 10^6\,\text{N/m}^2$$

Substituting this value into our solution for part (a) reveals

$$r = \sqrt{\frac{10\,000}{\pi \times 100 \times 10^6}}$$
$$= 5.64 \times 10^{-3} \text{ (or 5.64 mm)}$$

Example 9.10   *A 75 mm long steel rod, of diameter 11.28 mm, is subjected to a compressive stress of 25 kN. If the modulus of elasticity* E = 200 GN/m² *determine*

(a) the direct stress in the rod
(b) the change in length of the rod

SOLUTION

(a)   Unlike Example 9.9, this rod is in compression hence the force applied is

$$F = -25 \times 10^3 \text{ N}$$

and the direct stress is given by

$$\sigma = \frac{F}{A}$$
$$= \frac{-25\,000}{\pi(0.005\,64)^2}$$
$$= -250.2 \text{ MN/m}^2$$

(b)   The solution is identical to part (b) of Example 9.9.

$$E = \frac{\sigma}{\varepsilon}$$

or

$$\varepsilon = \frac{\sigma}{E}$$

but we also know that

$$\varepsilon = \frac{\Delta l}{l}$$

When rearranged for Δl, this gives

$$\Delta l = \varepsilon l$$

hence we obtain a general expression for Δl:

$$\Delta l = \frac{\sigma}{E} l$$

and therefore

$$\Delta l = \frac{-250.2 \times 10^6}{200 \times 10^9} 0.075$$
$$= -93.8 \times 10^{-6} \text{ (or } -0.0938 \text{ mm)}$$

Note that the − sign indicates the rod decreases in length.

Summary

Fundamental definitions

Direct stress: $\sigma = \dfrac{\text{force}}{\text{cross-sectional area}} = \dfrac{F}{A}$ (N/m^2)

Direct strain: $\varepsilon = \dfrac{\text{change in size}}{\text{original size}} = \dfrac{\Delta l}{l}$ (dimensionless)

Thermal strain: $\varepsilon = \alpha \Delta T$

Shear stress: $\tau = \dfrac{\text{shear force}}{\text{cross-sectional area}} = \dfrac{F}{A}$ (N/m^2)

Shear strain: $\gamma = \dfrac{x}{l} \approx \phi$ (dimensionless)

Material properties

Material properties are depicted on direct stress–strain curves or shear stress–strain curves.

The curves are obtained from standard tensile tests.

The following properties are important:

Modulus of elasticity: $E = \dfrac{\text{direct stress}}{\text{direct strain}} = \dfrac{\sigma}{\varepsilon}$ (N/m^2)

Yield stress σ_y and 0.1% proof stress $\sigma_{0.1\%}$

Modulus of rigidity: $G = \dfrac{\text{shear stress}}{\text{shear strain}} = \dfrac{\tau}{\gamma}$ (N/m^2)

A brittle material fails mechanically with little or no plastic deformation.
A ductile material fails mechanically after substantial plastic deformation has occurred.

Problems

Stress and strain relationships

9.1   A component in a car's suspension is subjected to a tensile force of 20 kN. The component is manufactured from hollow steel tube of external diameter 25 mm and wall thickness 4 mm, and whose modulus of elasticity $E = 209\,\text{GN/m}^2$. Determine
   (a)   the cross-sectional area of the component
   (b)   the direct stress in the component
   (c)   the direct strain in the component

9.2   A section of a beam is subjected to a shear force of 1 kN. The beam is constructed from square-section aluminium, 25 mm × 25 mm, and $G = 26\,\text{GN/m}^2$. Determine
   (a)   the shear stress in the beam
   (b)   the shear strain in the beam

9.3   A 2 m long section of railway track had a cross-sectional area of 2400 mm^2, measured on a day when the temperature was 15 °C. If the track is then laid on a day when the ambient temperature is 30 °C, determine
   (a)   the themal strain
   (b)   the change in length of the track
   Assume $\alpha = 12 \times 10^{-6}\,\varepsilon/\text{K}$

9.4 A 1.41 m long spar in a framework is subjected to a tensile load of 20 kN. The spar is constructed from aluminium tubing of internal diameter 40 mm and external diameter 50 mm. Assuming that $E = 70 \, \text{GN/m}^2$, determine
(a) the direct stress in the spar
(b) the direct strain in the spar
(c) the extension of the spar due to the applied load

9.5 A steel shaft of diameter 20 mm and length 2 m is subjected to a compressive load of 400 N. Determine
(a) the direct stress in the shaft
(b) the change in length of the shaft when loaded

9.6 A 15 kg steel ring of internal diameter 100 mm is to be fitted on to a steel shaft of similar external diameter. To ensure tightness of fit, the shaft has been manufactured such that its diameter is 100.075 mm. All dimensions have been checked at a temperature of 23 °C. Determine
(a) the temperature the ring should be raised to so that the ring can be slid on to the shaft
(b) the amount of heat required to achieve the temperature rise found in (a)

9.7 A replacement guide for the valves of an internal combustion engine cylinder-head has an external diameter of 18 mm. The cylinder-head is machined such that the replacement guide is fitted into a hole of diameter 17.95 mm. The cylinder-head and the guide have similar thermal coefficients of expansion, $\alpha = 12 \times 10^{-6} \, \varepsilon/\text{K}$. All machining was carried out at an ambient temperature of 18 °C. Determine
(a) the temperature change of the cylinder-head required to enable a sliding fit
(b) the temperature change of the replacement guide required to enable a sliding fit

9.8 A robotic arm uses a 250 mm long, 3 mm diameter steel wire as an actuator cable. In order to minimise positional errors, the extension of this cable must be less than 0.25 mm. Determine the maximum permissible load.

Material properties

9.9 Table P9.9 gives force–extension data for a 100 mm² tensile test specimen. The gauge length was 60 mm. Draw a graph of stress versus strain for the specimen then determine the modulus of elasticity and the 0.1% proof stress for the material.

Table P9.9

Applied load (kN)	Extension (μm)
13	20
26	40
39	60
52	80
57	90
61	100
63	110
65	150
66	200
67	250
66	300
60	350

9.10   A shaft of diameter 40 mm and length 2 m is subjected to a tensile load F. The 0.1% proof stress for the material is $250 \, MN/m^2$ and the modulus of elasticity is $200 \, GN/m^2$. Determine
    (a)   the maximum load the shaft can carry
    (b)   the change in length of the shaft when loaded to half the value determined in (a)

9.11   A supporting spar carries a tensile load of 25 kN when fully loaded. The spar is constructed from an aluminium alloy tube of yield strength $320 \, MN/m^2$. The tube has external diameter 50 mm and wall thickness 5 mm. Determine
    (a)   the direct stress the spar carries
    (b)   the factor of safety for the spar

9.12   Which of the following statements describes the properties of a ductile material?
    (a)   The material, when loaded to failure, exhibits little plastic deformation.
    (b)   A fractured tensile test specimen, when loaded to destruction, cannot be put back together.
    (c)   The stress–strain curve always exhibits clear linear and plastic regions.
    (d)   A ductile material is easily deformed.
    (e)   A ductile material has no modulus of elasticity.

9.13   Which of the following statements are applicable to a material obeying Hooke's law?
    (a)   Deflection is proportional to applied load.
    (b)   There is no elastic region for the material.
    (c)   Stress is not proportional to strain.
    (d)   For a given increase in stress, the increase in strain is exactly the same.

9.14   A shear-pin in a trailer is subject to a shear force of 5 kN. The shear-pin is to be made form an alloy steel whose yield stress is $350 \, MN/m^2$. If the material is to operate under a safety factor of 2.5, determine the minimum diameter of the pin which would satisfy this criterion.

9.15   A guillotine is to cut sheet steel of width 1 m and thickness 5 mm. If the blade shears the entire width in one pass, determine the minimum force required to achieve a successful shear. The maximum shear stress for the steel sheet is $80 \, MN/m^2$.

9.16   A 150 mm long bar is fixed at both ends. The bar is 25 mm × 25 mm square and manufactured from a steel alloy. Under operating conditions the bar is heated from a nominal 15 °C to an operating temperature of 45 °C. Assuming the fixings do not allow expansion, determine
    (a)   the direct stress under which the bar operates
    (b)   the reaction force induced by the fixings
You may assume that $\alpha = 12 \times 10^{-6} \, \varepsilon/K$ (*Hint*: when a component is subject to thermal expansion, thermal stress only exists when it is restrained.)

Chapter 10

Case studies

In this chapter we shall be examining the application of the tools we met earlier to real-life problems. Some of these problems I have encountered over the years, others have been donated by industry. In all cases the emphasis is placed on building a model.

10.1 Socket T-bar

A common tool in a mechanic's toolbox is the socket set. This set usually comprises a number of steel alloy sockets, which are used to provide a positive location on a nut or bolt. To turn the socket, the mechanic can use a number of devices, in this case study we shall examine the T-bar. Figure 10.1 illustrates a common T-bar design. In the worst case the bar is fully extended, as illustrated in Figure 10.1(a); a more efficient use is demonstrated in Figure 10.1(b). In this case study we shall examine the

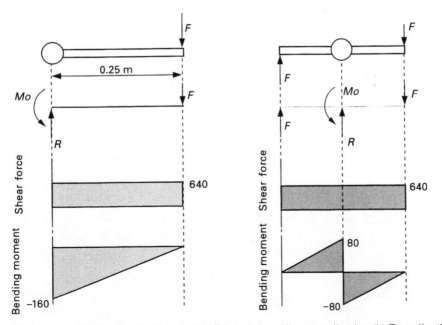

Figure 10.1 Applied loading to a socket T-bar: (a) cantilever application (b) T-application

reasons why this is so. Furthermore, we shall be selecting a suitable material for the bar, based on our calculations.

Socket sets are commonly used to tighten nuts to a specific 'torque', or to loosen nuts which were previously tight. The tightness of a nut or bolt is measured using a torque wrench. What the torque wrench actually measures is the *reaction moment* due to friction between the nut, bolt and component. So in reality we do not apply a torque to the wrench, we apply a moment. The moment is generated by applying a force F at a distance d. The function of the T-bar is to provide the distance d.

In this exercise the maximum torque (or moment) applied to tighten the nut is assumed to be 160 N m. Two ways to achieve this are demonstrated in Figure 10.1(a) and (b). In both cases a free-body diagram of the T-bar is shown. The reaction forces are the forces that the nut exerts upon the spanner.

Case A: Bar fully extended, Figure 10.1(a)
The moment generated by applying the force F in this configuration is

$$M = Fd$$
$$160 = F \times 0.25$$

hence

$$F = 640 \, \text{N}$$

To determine the reactions to the applied force, we need to consider the T-bar as a beam. By inspection we recognise that this system is effectively a cantilever beam (see Chapter 8). Hence the reaction moment is $Mo = 160 \, \text{N m}$ and the reaction force is $R = 640 \, \text{N}$.

Case B: Bar midspan, Figure 10.5(b)
In this case the applied moment generated by applying two equal and opposite forces to the T-bar is given by

$$M = F_1 d + F_2 d$$
$$160 = (F_1 \times 0.125) + (F_2 \times 0.125)$$
$$= 0.125(F_1 + F_2)$$

The system is symmetrical, $F_1 = F_2 = F$, so

$$160 = 0.125 \times 2F$$
$$F = 640 \, N$$

The reaction moment for this configuration is the applied moment, 160 N m. The reaction force is determined from Newton's second law:

$$\Sigma F - 0$$
$$640 + (-640) + R = 0$$

hence

$$R = 0$$

So immediately we can see that the tightening method of Case B (Figure 10.1(b)), applies no force to the nut or bolt. This reduces stresses in the system. The next time you tighten a wheel-nut on your car, try these two methods out. You will notice that the method shown in Figure 10.1(b) tightens the nut but does not move the car as well!

The shear force and bending moment diagrams may also be drawn for both cases. They are also illustrated in Figure 10.1(a) and (b). You should try and prove them for yourself. Note that in both cases the maximum shear forces are identical but the maximum bending moment is halved for case 10.1(b).

We can determine the shear stresses induced in the T-bar. Recalling the definition for shear stress (Chapter 9)

$$\tau = \frac{\text{shear force}}{\text{cross-sectional area}}$$
$$= \frac{640}{\pi \times 0.006^2}$$
$$= 5.7 \, \text{MN/m}^2$$

This tells us that, to withstand the shear forces, the T-bar's material must be able to withstand shear stresses of $5.7 \, \text{MN/m}^2$. From Chapter 9 we also know that an approximation for the yield stress of the material is twice this value, $\sigma_y = 11.4 \, \text{MN/m}^2$. Just to be sure, engineers also apply safety factors; a safety factor of 2 suggests that the material's yield stress (or 0.1% proof stress) should be no less than $\sigma_y = 22.8 \, \text{MN/m}^2$.

Note that we have not calculated all possible stresses. In particular, we have not considered two common force applicators, the foot and the lump hammer!

10.2 Hydroelectric power stations

Figure 10.2 illustrates a typical layout for a hydroelectric power station. A large water store, commonly a natural lake or reservoir, is many metres above the

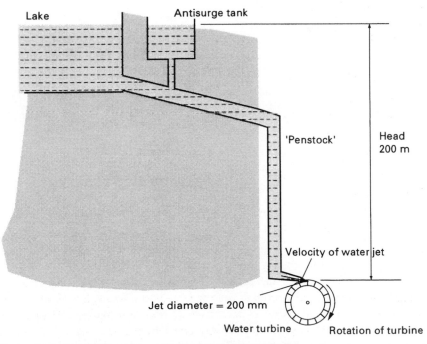

Figure 10.2 Typical layout for a hydroelectric power station

actual power station. To exit the lake, the water has to fall through pipework and pass through the powerstation; it then goes on its way to the sea. But how does this help engineers to produce electricity? Let us consider the energy of the system.

When the water is in the top reservoir, a lake in this instance, it has gravitational potential energy. From Chapter 8, we know that Bernoulli's equation defines this in terms of *head*. The water head is therefore the difference in height from the *last free surface* to the jet exit. There is often an antisurge tank in the pipeline which connects the lake to the penstock. This exists to smooth any fluctuations in flow which could damage the pipework below.

As the water falls down the penstock, under gravitational acceleration, the potential energy is transferred into kinetic energy and hence a velocity of flow. In Chapter 8 this was shown to have a maximum of $\sqrt{2gh}$ m/s. Hence the maximum velocity of flow is

$$v = \sqrt{2 \times 9.81 \times 200}$$
$$= 62.6 \, \text{m/s}$$

Now all the engineer has to do is transfer all of this kinetic energy into useful work in a generator. This is carried out using turbines. There are several turbines available to the engineer. A common design is the *Pelton wheel*, an *impulse turbine*.

Because we know the jet diameter, 0.2 m, we therefore know the maximum mass flow rate of water:

$$\dot{m} = \rho A v$$
$$= 1000 \times \pi \times 0.1^2 \times 62.6$$
$$= 1966.6 \, \text{kg/s}$$

The maximum power available is easily determined, since power is the rate of change of energy. In this case it is easier to think of the amount of energy passing through the station every second. Kinetic energy is defined as

$$E_k = \tfrac{1}{2} m v^2$$

Power is defined as the rate of energy, hence

$$P = \frac{dE_k}{dt} = \frac{d}{dt}(mv^2/2)$$

If velocity of flow is assumed to be constant, this becomes

$$P = \frac{1}{2} v^2 \frac{dm}{dt} = \frac{1}{2} \dot{m} v^2$$
$$= \frac{1}{2} \times 1966.6 \times 62.6^2$$
$$= 3.85 \, \text{MW}$$

This value is far higher than the power we actually recover. Firstly, if all of this power is to be recovered, the turbines have to be 100% efficient. But about 80–90% is the best we could expect. Also we have assumed that all of the potential energy of the water is transferred to kinetic energy, something which you should now understand as being an unattainable goal!

10.3   The bypass jet engine

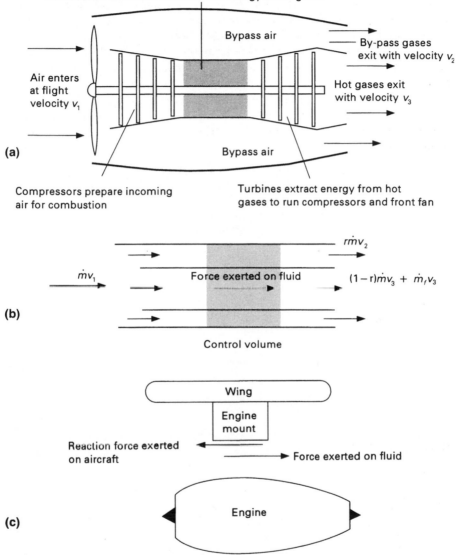

Combustion chambers increase the energy of the gases

Bypass air

By-pass gases exit with velocity v_2

Air enters at flight velocity v_1

Hot gases exit with velocity v_3

(a)

Bypass air

Compressors prepare incoming air for combustion

Turbines extract energy from hot gases to run compressors and front fan

$\dot{m}v_2$

$\dot{m}v_1$

Force exerted on fluid

$(1-r)\dot{m}v_3 + \dot{m}_f v_3$

(b)

Control volume

Wing

Engine mount

Reaction force exerted on aircraft

Force exerted on fluid

Engine

(c)

Figure 10.3   Principle of a bypass jet engine: (a) schematic, (b) engine model and (c) free-body diagram

If you examine modern civilian aircraft, you will discover some amazing facts. Next time you get the opportunity, normally when boarding, look into the engine itself. You will notice that it is a big hole with a large 'fan' on the front and a small engine inside. Why is the whole engine cowling not completely filled with engine? Let us examine it in more detail. Figure 10.3(a) illustrates a typical modern civil jet engine. The term *jet* arises because the motive force produced comes from the jet of gases emitted at the rear. At the front of the engine there is a large fan, which 'draws' in air from the outside. Part of the air goes into the engine, and the rest is transported *around* the engine from front to rear. The portion drawn into the engine is compressed (similar to a car's internal combustion engine) into the combustion chamber, where the real business occurs. Fuel enters the combustion chamber. It mixes with the compressed air and is ignited. This volatile combustion process produces gas temperatures of about 2200 °C and pressures in the region of 4 MN/m². These very hot gases are directed through *high-pressure turbines* which convert the stored energy in the gases to work. This work drives generators and pumps as well as the front fan, which is used to draw in the air in the first place! The gases are exhausted through the rear of the engine with the bypassed air.

To analyse the system we can examine a simple schematic model (Figure 10.3(b)), based around a *control volume*. The control volume ignores the internal workings of the engine and only considers 'what goes in' and 'what comes out'. The boundary conditions are also illustrated on the figure. For simplicity we assume the pressure at inlet and outlet are the same. The free-body diagram for the engine is illustrated in Figure 10.3(c). Recall from Chapter 7 that the force required to change the momentum of a fluid is given by

$$F = \dot{P} = \dot{m}(v_2 - v_1)$$

We must develop this further, but we must not forget the bypass ratio r and the mass flow rate of fuel injected into the combustion chamber $\dot{m}_f$. We can therefore generate an equation for the force acting on the fluid (which is equivalent to the thrust of the engine):

$$F = [(\dot{p}_{\text{bypass air}}) + (\dot{p}_{\text{engine air}}) + (\dot{p}_{\text{fuel}})] - (\dot{p}_{\text{entry air}})$$
$$F = [(r\dot{m}v_2 + (1 - r)\dot{m}v_3) + (\dot{m}_f v_3)] - (\dot{m}v_1)$$

(10.1)

Furthermore, we can examine the change in kinetic energy of the gases. From the previous case study we saw that the power (or energy per unit time) of a fluid jet is given by

$$P = \tfrac{1}{2}\dot{m}v^2$$

At intake, the energy per unit time of the air is given by

$$P_1 = \tfrac{1}{2}\dot{m}v_1^2$$

(10.2)

On exit, the energy per unit time of the gases is

$$P_e = \tfrac{1}{2}r\dot{m}v_2^2 + \tfrac{1}{2}(1 - r)\dot{m}v_3^2 + \tfrac{1}{2}\dot{m}_f v_3^2$$

(10.3)

Let us consider what happens to these energies if the mass flow rate of the inlet air is kept constant but the bypass ratio changes. To simplify the analysis we will assume that the contribution of the fuel's kinetic energy to the total exit energy is small in comparison to the air. Engineers like to define the operation of a machine by efficiency. One simple way of defining efficiency is by examining a ratio of the total work to useful work which has been done. This may be defined as

$$\text{Percentage efficiency } (\eta) = \frac{\text{useful work per unit time}}{\text{total work per unit time}} \times 100$$

where the useful work per unit time is that done by the jet, $P = Fv$. The total work per unit time is the work done by the jet *plus* the work per unit time on the engine's gases, $(P_e - P_i)$. Hence

$$\eta = \frac{Fv}{Fv + (P_e - P_i)} \times 100$$

where F is obtained from equation (10.1); P_i and P_e are obtained from (10.2) and (10.3) respectively.

Before we can analyse this further, we need to consider some typical design data for a modern jet engine. A typical set of operating conditions may be

$$\text{entry velocity, } v_1 = 172 \, \text{m/s}$$
$$\text{entry air mass flow rate, } \dot{m} = 268 \, \text{kg/s}$$
$$\text{exit velocity of bypassed air, } v_2 = 230 \, \text{m/s}$$
$$\text{engine thrust, } F = 49 \, \text{kN}$$

Figure 10.4 shows a graph of the percentage efficiency as the bypass ratio changes from 0 to 25% ($r = 0–0.25$). As the bypass ratio changes, the efficiency reaches a plateau. The optimum bypass ratio is clearly demonstrated for this particular operating condition, $r = 16.7\%$. This simple model demonstrates that a bypass engine can be more efficient than a straight jet engine. If this is the case, why are bypass engines not used in all aircraft? Firstly this was only a simple model and does not take into account all variables, e.g. mass flow rate of the fuel. Secondly bypass engines are bulky by nature, so they may

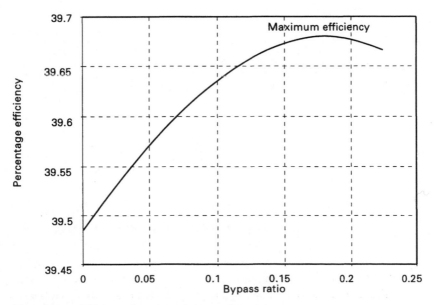

Figure 10.4 Efficiency for a bypass jet engine

not suit particular aircraft. Thirdly fluid dynamics changes when you consider supersonic flight (faster than the speed of sound)! If you study this topic further, all may be revealed!

10.4   *Hydraulic actuator*

Data courtesy of Denison Mayes Group

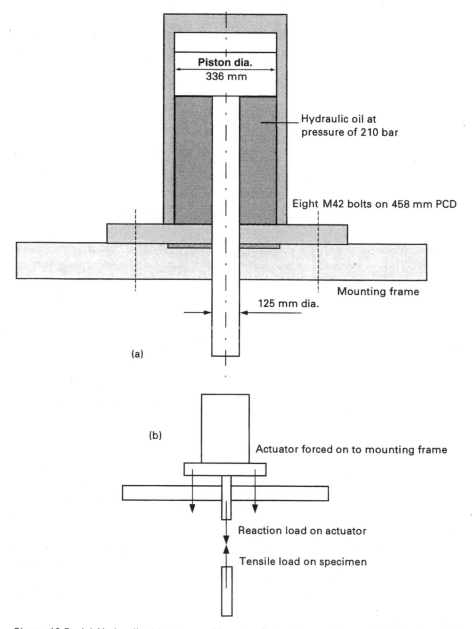

Figure 10.5   (a) Hydraulic actuator used in a tensile testing machine and (b) the free-body diagram for its housing

The tensile testing machine discussed in Chapter 9 relies upon a hydraulic ram to produce the tensile forces required to break a tensile test specimen. Figure 10.5(a) illustrates a hydraulic actuator which could be used in this machine. The internal chamber is pressurised by high-pressure oil (up to 210 bar) which enters the actuator via ports in the top boss. Clearly, at these pressures, design engineers have to ensure safety. The sort of question they would ask – and this is commonly a part of a failure modes and effects analysis (FMEA) – would be, How would this unit fail?

One mode of failure would be tensile failure of the actuator shaft itself. First we need to determine the maximum force on the actuator rod. Recalling hydrostatic pressure forces from Chapter 7, we note that

$$p = \frac{F}{A}$$

or

$$F = pA$$
$$= 210 \times 10^5 \times \pi(0.168^2)$$
$$= 1862 \, kN$$

which is the maximum tensile load the actuator must withstand. In real life the piston diameter is often slightly less than the bore diameter. This is to allow for seals to be fitted, as we do not want pressurised oil leaking between the walls of the cylinder and the piston.

Recalling the direct stress calculations from Chapter 9

$$\sigma = \frac{F}{A}$$
$$= \frac{1862 \times 10^3}{\pi(0.062 \, 5)^2}$$
$$= 151.7 \, MN/m^2$$

Which, if we are to allow a safety factor of 2, suggests that we require a material that can withstand a tensile stress of 303 MN/m^2.

Another failure mode is for the body to be forced off the machine. For this to occur, the eight M42 bolts would have to fail in tension. But an examination of the free-body diagram for the actuator housing (Figure 10.5(b)) shows that the positioning of the actuator body has relieved the bolts of any loads when the system acts in tension. If the piston applies a tensile force, the body is subjected to an equal and opposite force. This force has the effect of forcing the housing on to the main body rather than pulling it off!

If the force is in the other direction, through misuse, then the eight M42 bolts have to support the full 1862 kN. Since they are symmetrically mounted on a 485 mm *pitch circle diameter*, they must all carry the same load

$$F = \frac{1862}{8}$$
$$= 232.75 \, kN$$

The load is axial and tensile, hence the bolts are under direct stress

$$\sigma = \frac{F}{A}$$
$$= \frac{232.75 \times 10^3}{\pi(0.021^2)}$$
$$= 168 \, \text{MN/m}^2$$

If a safety factor of 2 is applied, this suggests that the bolts should be able to withstand at least $336 \, \text{MN/m}^2$. One way of limiting this stress is to restrict the pressure of the fluid to the return stroke of the piston.

The final item we can consider here is the quantity of oil required for full travel of the piston. The stroke is 250 mm – *stroke* is an engineering term for total distance travelled by a component which moves back and forth – hence the *swept volume* is given by

$$V = Al$$
$$= \pi(0.168^2) \times 0.25$$
$$= 0.022 \, \text{m}^3$$

If the maximum velocity of the piston is to be 0.025 m/s, this yields a maximum volumetric flow rate of

$$\dot{V} = Av$$
$$= \pi(0.168^2) \times 0.025$$
$$= 0.0022 \, \text{m}^3/\text{s or } 2.2 \, \text{litres/s}$$

10.5   Drive ratio for a racing lawnmower

This may seem bizarre, but the author has spent many a long winter building lawnmowers suitable for racing in the summer months, in particular, the gruelling annual 12 hour race! Unfortunately the team never actually won, but we had a lot of fun competing, which is the main thing after all. In this case study we shall be selecting gear ratios for a selected speed. Furthermore, we shall be examining the stress induced in a key which is used to locate the drive-wheels.

The main motive power was a standard, 6 kW (at 3600 rev/min) petrol internal combustion engine. For the 12 hour race, a nominal speed of 60 km/h at an engine speed of 3600 rev/min was desirable. The drive-wheels were 200 mm external diameter and the rear axle used a small gearbox which had a reduction of 2:1. The engine was connected to the gearbox via a drive-belt which ran on two pulleys, as illustrated in Figure 10.6(a). The pulley on the engine was 150 mm in diameter. The rear axle was constructed from a solid 25 mm steel shaft, and the wheels were keyed on to the shaft as illustrated in Figure 10.6(b).

The torque transmitted at the crankshaft may be determined from

$$P = T\omega$$

which when rearranged for T becomes

$$T = \frac{P}{\omega}$$

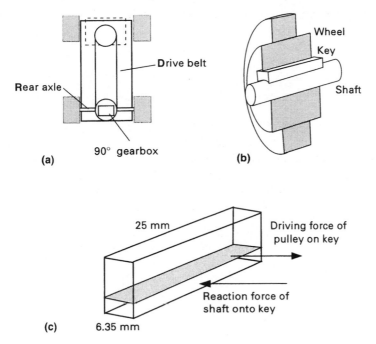

(a)

90° gearbox

Rear axle

Drive belt

(b)

Wheel

Key

Shaft

(c)

25 mm

6.35 mm

Driving force of
pulley on key

Reaction force of
shaft onto key

Figure 10.6 Racing lawnmower: (a) schematic, (b) wheel-hub schematic and (c) shear force acting on key

hence

$$T = \frac{6000}{120\,\pi}$$
$$= 15.92\,\text{N m}$$

To determine the torque at the wheels, we need to determine the angular velocity of the wheels. Energy must be conserved. Power is transmitted by the drive-belt. Hence if we assume no losses, the power at the wheels must be the same as at the engine.

The desired speed is 60 km/h, and the wheel diameter is 200 mm. Recalling peripheral velocity from Chapter 2, we can determine the angular velocity of the wheels from

$$v = \omega r$$

which when rearranged becomes

$$\omega = \frac{v}{r}$$

hence

$$\omega = \frac{(60/3.6)}{0.1}$$
$$= 166.67\,\text{rad/s}$$

Therefore, the torque transmitted to the wheels is

$$T = \frac{6000}{166.67}$$
$$= 36\,\text{N m}$$

This is greater than the torque at the crankshaft. Thus the key has to withstand a shear force of

$$F = \frac{T}{r}$$
$$= \frac{36}{0.012\,5}$$
$$= 2.88\,\text{kN}$$

Figure 10.6(c) illustrates the key operating under shear. The key length is 25 mm and its width is 6.35 mm. Hence the cross-sectional area which carries the shear is

$$A = 0.025 \times 0.006\,35$$
$$= 0.000\,159\,\text{m}^2$$

Thus the shear stress carried by the key is

$$\tau = \frac{F}{A}$$
$$= \frac{2880}{0.000\,159}$$
$$= 18.1\,\text{MN/m}^2$$

which suggests (from Chapter 9) that a mild steel key would more than suffice.

To determine the overall drive ratio, R_{we}, the speed of the wheels should be compared with that of the engine:

$$R_{we} = \frac{\omega_{wheels}}{\omega_{engine}}$$
$$= \frac{166.67}{120\pi}$$
$$= 0.44$$

This tells us that the wheels have to have an angular velocity which is 0.44 of the engine's angular velocity. Unfortunately this does not help us in relation to pulley sizes because of the gearbox. Since the gearbox already has a ratio of 2:1 (or 0.5), the belt drive ratio, which is determined by pulley size, has to alter this ratio to 0.44. The drive ratio of the engine to the gearbox can be defined as

$$R_{eg} = \frac{\omega_{engine}}{\omega_{gearbox}}$$

or

$$\omega_{gearbox} = \frac{\omega_{engine}}{R_{eg}}$$

The drive ratio of the gearbox to the wheels is

$$R_{gw} = \frac{\omega_{gearbox}}{\omega_{wheels}}$$

Into which we can substitute our relationship for the gearbox–engine ratio:

$$R_{gw} = \frac{\omega_{engine}}{\omega_{wheels} \times R_{eg}}$$

This can be rearranged to yield an expression for the ratio of wheel speed to engine speed:

$$R_{ew} = \frac{\omega_{engine}}{\omega_{wheels}} = R_{gw} \times R_{eg}$$

Substituting in our known values gives

$$0.44 = 0.5 \times R_{eg}$$

hence

$$R_{eg} = 0.88$$

Recalling from Chapter 2 that gearing ratio was defined by

$$\frac{\omega_1}{\omega_2} = \frac{D_2}{D_1}$$

where ω_1/ω_2 is the ratio of engine speed to gearbox speed R_{eg}, hence

$$0.88 = \frac{150}{D_1}$$

thus

$$D_1 = \frac{150}{0.88}$$
$$= 170 \, mm$$

This is the diameter of the front pulley required to achieve the correct speed. The effect of the ratios on component speed is illustrated in Figure 10.7.

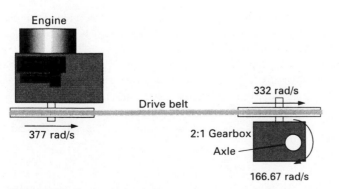

Figure 10.7 Engine/drive belt

10.6   Coastal lifts

Lift systems are often used at coastal resorts to transport people up and down cliffs, as illustrated in Figure 10.8. Two cars ride on parallel tracks which are laid up the cliff face. As one car moves up the cliff the other comes down. The two cars are joined together by cables which run around a large pulley at the top of the cliff. In effect, the cable utilises gravity to full advantage; the car moving down the cliff 'pulls' the car moving up the cliff. In this case study we shall be examining the forces which cause the lifts to move up and down the cliff face.

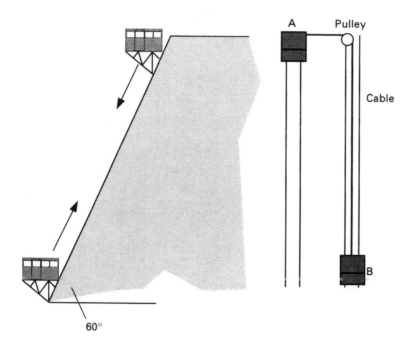

Figure 10.8   Schematic of a cliff lift

Consider the situtation where car A is at the top of the cliff and is loaded such that its mass is 1000 kg, and the car at the bottom of the cliff is loaded such that its mass is 750 kg. What is the acceleration of the cars?

Assuming that the pulleys at the top of the slope are frictionless, the tension within the cable will be constant. Drawing a free-body diagram for car A (Figure 10.9(a)) reveals the tension in the cable. Note that we have oriented the x and y axes with the cliff face to simplify our calculations. Applying Newton's second law to the component of the body force acting down the slope and in line with the cable tension, $mg \cos(-150°)$, yields

$$\Sigma F_x = m\ddot{x}$$

$$[(1000 \times 9.81) \cos(-150°)] + F_T = 1000\ddot{x} \tag{10.1}$$

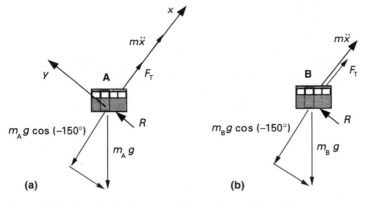

Figure 10.9 Free-body diagrams: (a) car A and (b) car B

In the y-direction the sin component of mg is supported by the reaction force R. Considering the free-body diagram of car B (Figure 10.9(b)) reveals that in the x-direction we have

$$\Sigma F_x = m\ddot{x}$$

$$[(750 \times 9.81) \cos (-150°)] + F_T = 750\ddot{x}$$

(10.2)

However, in both (10.1) and (10.2) $\ddot{x}$ must be equal and opposite. We can therefore rearrange them to give expressions for $\ddot{x}$ and subsequently equate them:

$$\ddot{x} = (-6.372 + F_T/750) = -(-8.496 + F_T/1000)$$

from which we find F_T

$$-6372 + 1.33F_T = 8496 - F_T$$
$$2.33F_T = 14\,868$$
$$F_T = 6381 \text{ N}$$

Hence we can find the acceleration of the cars from either (10.1) or (10.2).

From (10.1)

$$\ddot{x}_A = (-8.496 + 6381/1000) = -2.1 \text{ m/s}^2$$

From (10.2)

$$\ddot{x}_B = (-6.372 + 6381/750) = 2.1 \text{ m/s}^2$$

Hence car A accelerates down the cliff and B accelerates up the cliff.

Appendices

Standard trigonometric functions

Right-angled triangle

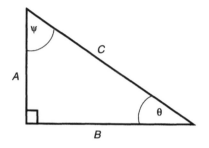

$$\sin \theta = A/C \qquad \cos \theta = B/C \qquad \tan \theta = A/B \qquad \psi = 90 - \theta$$
$$C^2 = A^2 + B^2$$

Any triangle

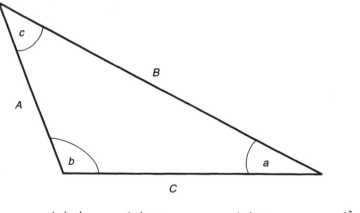

$$B = \frac{A \sin b}{\sin a}; \quad C = \frac{A \sin c}{\sin a}; \quad \tan a = \frac{A \sin c}{B - (A \cos c)}; \quad \cos a = \frac{b^2 + c^2 - a^2}{2\,BC}$$

$$a + b + c = 180°: \text{Area} = \frac{AB \sin c}{2}$$

Double-angle formulae

$$\sin(a+b) = \sin a \cos b + \sin b \cos a$$
$$\sin(a-b) = \sin a \cos b - \sin b \cos a$$
$$\cos(a+b) = \cos a \cos b - \sin a \sin b$$
$$\cos(a-b) = \cos a \cos b + \sin a \sin b$$
$$\tan(a+b) = \frac{\tan a + \tan b}{1 - \tan a \tan b}$$
$$\tan(a-b) = \frac{\tan a - \tan b}{1 + \tan a \tan b}$$

Other relationships

$$\sin^2 \theta + \cos^2 \theta = 1$$
$$\sin 2\theta = 2 \sin \theta \cos \theta$$
$$\cos 2\theta = 2 \cos^2 \theta - 1$$

Standard differentials and integrals

$f(x)$	$\dfrac{df(x)}{dx}$	$\int f(x)dx$
x	1	$x^2/2$
Ax^n	$nA\,x^{n-1}$	$\dfrac{Ax^{n+1}}{n+1}$
$\cos x$	$-\sin x$	$\sin x$
$\sin x$	$\cos x$	$-\cos x$
$\tan x$	$\dfrac{1}{\cos^2 x}$	$-\ln(\cos x)$
$\cos Ax$	$-A\sin Ax$	$\dfrac{\sin Ax}{A}$
$\sin Ax$	$A\cos Ax$	$-\dfrac{\cos Ax}{A}$
e^{nx}	ne^{nx}	$\dfrac{e^{nx}}{n}$

Numerical differentiation and integration

Consider the curve depicted in Figure C.1. The slope of the curve at x_0 can be determined using 'three point' formulae, e.g.

$$\frac{df(x)_0}{dx} \approx \frac{1}{2h}(-3f_0 + 4f_1 - f_2)$$

$$\frac{df(x)_2}{dx} \approx \frac{1}{2h}(f_0 - 4f_1 + 3f_2)$$

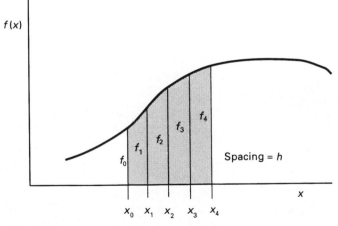

$f(x)$

f_4

f_3

f_2

f_1

f_0

Spacing = h

x

$x_0 \quad x_1 \quad x_2 \quad x_3 \quad x_4$

Figure C.1

However, numerical differentiation is particularly susceptible to noise and should be avoided if possible. Numerical integration is less susceptible to noise, and a common method is Simpson's rule.

$$\int_{x_0}^{x_n} f(x)dx \approx \frac{h}{3}(f_0 + 4f_1 + 2f_2 + \ldots + 2f_{n-2} + 4f_{n-1} + f_n)$$

Note that there must be an *odd* number of points. To obtain the area under the graph above between x_0 and x_4 there are five points:

$$\int_{x_0}^{x_n} f(x)dx \approx \frac{h}{3}(f_0 + 4f_1 + 2f_2 + 4f_3 + f_4)$$

For further details on numerical differentiation and numerical integration see, for example, *Advanced Engineering Mathematics*, E Kreyzig, J Wiley.

Gravitational acceleration

Newton's law of gravitation

Newton established that the force of attraction between two bodies due to gravity is given by

$$F = G\frac{m_1 m_2}{r^2} \tag{D.1}$$

Thus the acceleration due to gravity of a body in the earth's gravitational field is given by

$$g = \frac{Gm_e}{R^2} \tag{D.2}$$

where

G = the constant of gravitation = 6.673×10^{-11} m^3/kg s

m_e = mass of the earth = 5.976×10^{24} kg

R = radius from the centre of the earth

(the radius of the earth R_e is about 6.37×10^6 m)

g = the magnitude of a vector whose direction is towards the centre of the earth.

International gravity formula

Equation (D.2) does not allow for the rotation of the earth. The international gravity formula gives the value of g at any lattitude (at sea level):

$$g = 9.780\,327(1 + 0.005\,279\,\sin^2\gamma + 0.000\,023\,\sin^4\gamma + \ldots) \tag{D.3}$$

where γ is lattitude.

Questions selected from Staffordshire University examination papers

This appendix contains a number of examination questions which you should use for revision and examination practice. The questions should take no longer than 20 minutes each. You should be able to attempt all questions from memory. If you can do this, you will be well armed for future studies!

1. (a) Define direct stress.
    (b) Define direct strain.
    (c) A steel rod of diameter 20 mm is subjected to a tensile load of 8 kN. Determine
        (i) the direct stress which exists in the rod
        (ii) the direct strain which exists in the rod
        You may assume that Young's modulus of elasticity is $E = 200 \, GN/m^2$ for the rod material.

2. (a) Define Archimedes' principle.
    (b) A pontoon bridge is constructed from a number of pontoons tied together. Each individual pontoon is 1 m square, 4 m long and has a mass of 350 kg.
        (i) Determine the draught of each pontoon.
        (ii) If the pontoon is completely sealed, determine the maximum load a single pontoon can carry without sinking.

3. (a) Define the principle of conservation of energy.
    (b) Define the principle of conservation of momentum.
    (c) A jet aircraft of mass 1500 kg approaches a stationary aircraft carrier horizontally with a speed of 140 km/h. In an effort to stop the aircraft, a hook at the rear of the aircraft catches on to a cable stretched across the deck, once the aircraft touches the deck of the carrier. The cable is itself attached to springs having a total stiffness of 1389 N/m and the aircraft is 8 m long.
        (i) Determine the minimum length l of the aircraft carrier so that it accommodates the stopping distance of the aircraft.
        (ii) If no other form of energy dissipation exists, what will eventually happen to the aircraft after it has come to a halt?

4. Figure Q4 illustrates a typical belt-drive record-player turntable for single's. The normal speed of the turntable platter is 45 RPM (revolutions per minute), and the turntable is driven by a DC electric motor via a belt. The turntable acts as the large pulley, the belt running on its periphery. The turntable has a diameter of 300 mm.
   (a) Determine the angular velocity of the turntable in rad/s.
   (b) Determine the peripheral velocity of the turntable, and hence the linear velocity of the belt.
   (c) If the pulley mounted on the DC motor has a diameter of 30 mm, determine the running speed of the electric motor in RPM.

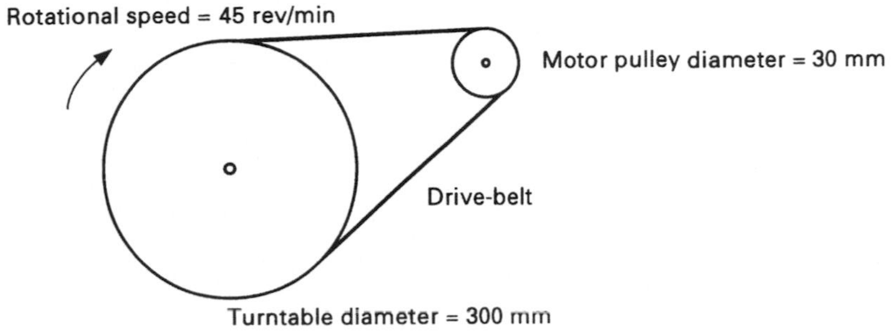

Rotational speed = 45 rev/min

Motor pulley diameter = 30 mm

Drive-belt

Turntable diameter = 300 mm

Figure Q4

5. (a) Describe the principle of conservation of energy.
   (b) Describe the principle of conservation of momentum.
   (c) A projectile of mass 0.09 kg approaches a stationary target horizontally with an initial velocity of 120 m/s. On impact the projectile is totally embedded in the target, the projectile and target subsequently acting as one body.
      (i) If the target mass is 10 kg, determine the velocity of the combined projectile and target after impact.
      (ii) Determine the kinetic energy before and after impact, describe the processes which take place during impact, and hence discuss the transfer of energy in the system.

6. (a) A cylinder (commonly called a bobbin) of diameter 1500 mm is used to store nylon twine, the twine being wrapped around the bobbin (as illustrated in Figure Q6). On demand, the bobbin can be transferred to a weaving machine

Speed of twine leaving bobbin = 30 m/s

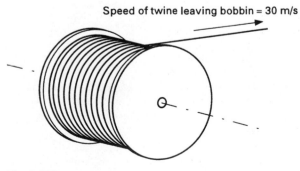

Figure Q6

and the nylon removed at a constant rate; this causes the bobbin to rotate at a constant speed. The velocity of the nylon twine leaving the bobbin is 30 m/s.
  (i)   Determine the angular velocity of the cylinder/bobbin.
  (ii) Determine the rotational speed (in rev/min) of the cylinder/bobbin.
(b) If the moment of inertia of the bobbin is 0.25 kg m², determine
  (i)   the angular momentum of the bobbin
  (ii) the kinetic energy of the bobbin

7.   (a) Given that Bernoulli's equation may be written as

$$\frac{p_1}{\rho g} + \frac{v_1^2}{2g} + z_1 = \frac{p_2}{\rho g} + \frac{v_2^2}{2g} + z_2$$

where

p = pressure
ρ = density
v = velocity
z = height
g = gravitational acceleration

subscript 1 indicates position 1
subscript 2 indicates position 2
Describe, with the aid of figures, what each term signifies.
(b) Figure Q7 illustrates a syphon utilised to empty a water storage tank. For the conditions illustrated, determine
  (i)   the velocity of flow at exit C
  (ii) the absolute pressure at point B
  (iii) the volumetric flow rate

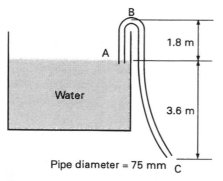

Figure Q7

8.   Figure Q8 illustrates the stress–strain curve obtained from a tensile test carried out on a specimen of normalised 0.1% carbon steel. The material was tested to failure.
(a) Indicate the following points or regions:
  (i)   elastic region
  (ii) plastic region
  (iii) ultimate stress

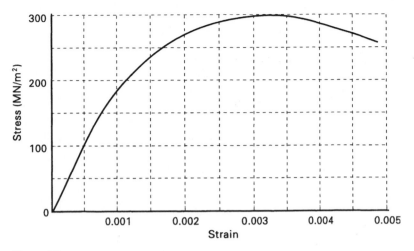

Figure Q8

    (b)   From Figure Q8 determine the following material properties:
        (i)   modulus of elasticity
        (ii)   0.1% proof stress

9.   A new design of battle tank is able to fire projectiles with a muzzle velocity of 240 m/s, the barrel of the gun being 3 m from ground level. On its first target practice, the target is placed 670 m away from the tank and is at the same level as the gun barrel. The tank commander aims the gun barrel horizontally at the target and fires.
    (a)   Sketch a diagram describing the path of the projectile, indicating salient points.
    (b)   Determine the time for which the projectile is in flight.
    (c)   Determine the horizontal distance by which the projectile misses the target.
    (d)   The bombardier suggests setting the gun at 3.28˙ to the horizontal; show that this is the correct setting for the required range of 670 m.
    (e)   Sketch the path of the projectile corresponding to the gun setting of part (d).

10.   (a)   State the three constant-acceleration equations which describe the relationship between uniform linear acceleration, velocity, distance and time.
     (b)   A sweet-wrapping machine 'fires' wrapped sweets horizontally from the wrapping-head into a storage container, as illustrated in Figure Q10. The sweets exit the wrapping head at a height of 200 mm from the top of the storage box and at a distance of 2 m from the nearest edge.
        (i)   Sketch the trajectory of the sweet from where it leaves the wrapping-head to where it enters the container. Indicate important points on the diagram.
        (ii)   Determine the minimum velocity each sweet must have attained at the wrapping-head for it to reach the container.
        (iii) Determine the maximum velocity of each sweet so that it does not overshoot the container.

Wrapping machine

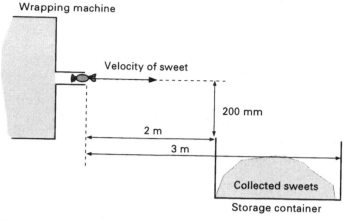

Figure Q10

11.   (a)   A joint in a framework used to support seating in an auditorium may be modelled as shown in Figure Q11(a). For the loading shown, determine the resultant force acting at the joint.

(b)   A friction joint is illustrated in Figure Q11(b). It is anticipated that this joint will replace the joint in Figure Q11(a). If the loading is the same as before, determine the friction coefficient required at the wall joint interface to ensure no sliding of the friction pad over the wall.

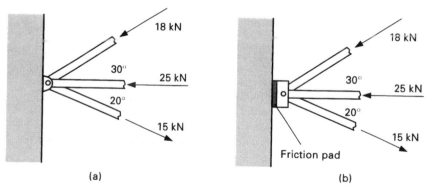

(a)   (b)

Figure Q11

12.   (a)   Convert the following values to absolute values:
      (i)   a gauge pressure of $5\,MN/m^2$ to absolute pressure
      (ii)   a gauge pressure of 40 mm of water to absolute pressure
      (iii) a temperature of $10\,^{\circ}C$ to absolute temperature

(b)   State the principle of continuity of flow.

(c)   A pipe network is illustrated in Figure Q12. If the flow of water into A is 45 kg/s, determine
      (i)   the mass flow rate of water exiting pipe C
      (ii)   the velocity of the flow at C

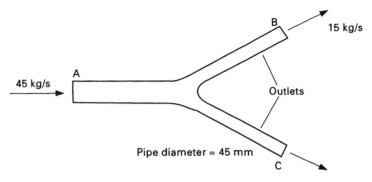

Figure Q12

13. (a) State the principle of conservation of energy.
    (b) A car of mass 2 tonnes is parked at the top of a hill, as illustrated in Figure Q13. If the car is allowed to roll down the hill freely, determine the maximum velocity the car can achieve at the base of the hill.
    (c) The car described in part (b) has a new design of front bumper which can deflect as a spring under impact. If the bumper has an effective stiffness of 800 MN/m, determine the maximum deflection of the bumper, assuming the car hits an object at the maximum velocity determined in (b).

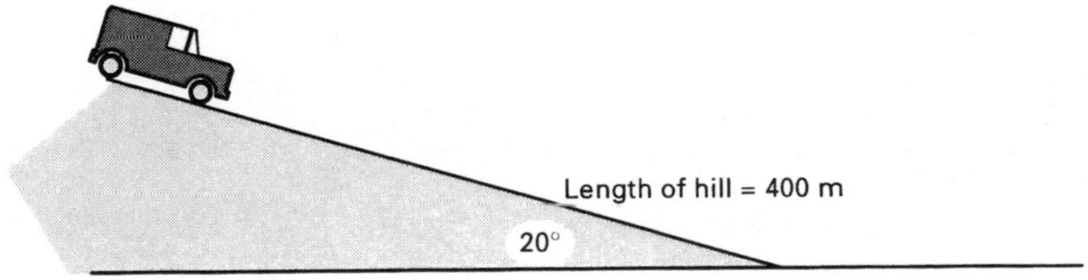

Length of hill = 400 m

20°

Figure Q13

14. (a) Define linear velocity.
    (b) Define linear acceleration.
    (c) Define angular velocity.
    (d) In a mercy mission, bags of grain of mass 52 kg are to be dropped from the rear of a transport plane flying at a height of 150 m with a ground speed of 380 km/h. The bags of grain are to be dropped into containers made from corrugated sheet steel to form a ring 20 m in diameter. Determine the maximum and minimum distances from the centre of the container at which the bags must be dropped such that they land in the container.

15.   Figure Q15 illustrates a simply supported beam which is subjected to three point
loads.

(a)   Show that the reaction forces at the left-hand and right-hand supports are

$$R_L = 375\,N$$
$$R_R = 625\,N$$

(b)   Draw a shear force diagram for the beam, indicating salient values.

(c)   Draw a bending moment diagram for the beam, indicating salient values.

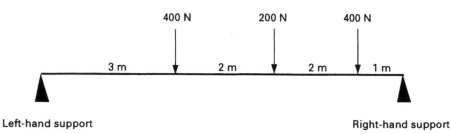

Left-hand support   Right-hand support

Figure Q15

Answers to end-of-chapter problems

Chapter 1

1.2 $A = (2.5, 7)$
     $B = (3, 4)$
     $C = (7, 3)$

1.3 $A_B = (-0.5, 3)$

1.4 $A_C = (-4.5, 4)$

1.6 $A_D = (-2, 4, -4)$
     $B_D = (1, 5, -1)$
     $C_D = (2, 0, 0)$

1.7 $OA = 2.5i + 7j$
     $OB = 3i + 4j$
     $OC = 7i + 3j$

1.8 $AC = 4.5i - 4j$

1.9 $d = 28i - 18j$

1.10 $a - b = 42i + 22j$
      $b - a = -42i - 22j$
      $c - a = -65i - 17j$

1.11 $a - b - c = 62i + 32j$

1.12 $|a| = 45.4$ $\angle a = 8.84°$
      $|b| = 15.3$ $\angle b = -78.7°$
      $|c| = 22.4$ $\angle c = -153.4°$

1.13 $ab = 10.9i + 43.7j$
      $ac = -40.9i + 50.5j$
      $ad = -86.97i - 86.97j$

1.14 $e = -117i + 7.23j$
      $= 117.2$ at $176.5°$

1.15 $a = 70$ at $90°$
      $b = 88.3$ at $-59.4°$
      $c = 58.3$ at $39.4°$

1.16 $2.56\,km$, $2.56 \times 10^3\,m$
      $45\,mm$, $45 \times 10^{-3}\,m$
      $2.35\,µm$, $2.35 \times 10^{-6}\,m$

1.17 $355.6 \times 10^{-3}\,m$
      $381\,mm$
      $76.2 \times 10^{-6}\,m$, $76.2 \times 10^{-3}\,mm$ and $76.2\,µm$

1.18 $10\,9\,kg$
      $92.2\,kg$

1.19 $261.3°$
      $0.798\,rad$
      $1170°$, $20.42\,rad$
      $26.74\,rad$

Chapter 2

2.2 (a) $v = 0\,m/s$
     (b) $v = -0.5\,m/s$
     (c) $v = 0.2\,m/s$

2.3 $41.67\,m/s$, $150\,km/h$; direction

2.4 $v_x = -1\,m/s$

2.5 $a = 1\,m/s^2$

2.6 $s(t) = t^2/2$
     $v(t) = t$
     $a(t) = 1$

2.7 $a = -9.81\,m/s^2$
     $+y$-axis is vertically upwards

2.8 $a = 0.97\,m/s^2$

2.9 $s = 75\,m$

2.10 (a) $a_x = 1\,\text{m/s}^2$
     (b) $a_y = -0.5\,\text{m/s}^2$
     (c) $a_x = 0\,\text{m/s}^2$
     (d) $a_y = -3\,\text{m/s}^2$

2.13 $\omega = 281\,\text{rad/s}$

2.14 (a) (i) $t = 2.8\,\text{s}$
         (ii) $t = 1.4\,\text{s}$
     (b) $v = 1.05\,\text{m/s}$ (low) and
         $1.57\,\text{m/s}$ (high)

2.15 (a) $\omega = 4.71\,\text{rad/s}$
     (b) $v = 0.72\,\text{m/s}$
     (c) $a_c = 3.38\,\text{m/s}^2$

2.16 (a) $\omega = 1.24 \times 10^{-3}\,\text{rad/s}$
     (b) $v = 7.9\,\text{km/s}$
     (c) $a = 9.79\,\text{m/s}^2$

2.17 (a) $\omega_1 = -104.7\,\text{rad/s}$
     (b) $v_1 = 5.24\,\text{m/s}$
     (c) $\omega_2 = 69.8\,\text{rad/s}$
         $= 667\,\text{rev/min}$

2.20 $v = 12.5\,\text{m/s}$
     $\omega = 62.5\,\text{rad/s}$

2.21 $x(t) = 0.025 \sin \pi t$
     $v(t) = 0.0785 \cos \pi t$
     $a(t) = -0.247 \sin \pi t$

2.22 $v = 0.94\,\text{m/s}$
     $a = 11.85\,\text{m/s}^2$

2.23 $v_B = 27.78\,\text{m/s}$ at $45°$

2.24 (a) $t = 58$ minutes and 3 seconds
     (b) $v_{BA} = -4.17\,\text{m/s}$

2.25 $v_{BA} = -101.39\,\text{m/s}$

2.26 $v_{AB} = 3.54\mathbf{i} - 8.54\mathbf{j}$
     $= 9.24\,\text{m/s}$ at $-67.5°$

Chapter 3

3.1 $v_1 = 60.25\,\text{m/s}$

3.2 (a) $t = 6.14\,\text{s}$
    (b) $t = 1.4\,\text{s}$ or $10.78\,\text{s}$

3.3 $t = 2.26\,\text{s}, v = 22.15\,\text{m/s}$

3.4 $h = 31.86\,\text{m}, t = 5.21\,\text{s}$

3.5 (a) $a = 1.3\,\text{m/s}^2$
    (b) $s = 583.3\,\text{m}$

3.6 $h = 326.2\,\text{m}, t = 8.15\,\text{s}$

3.7 $a = -6.94\,\text{m/s}^2, s = 69.5\,\text{m}$

3.8 $\alpha = -17.5\,\text{rad/s}^2$, 338.2 rev

3.9 $t = 20.6\,\text{s}$, 524 rev

3.10 460.5 rev

3.11 (b) $t = 3.19\,\text{s}$
     (c) $x = 47.9\,\text{m}$

3.12 (a) $t = 7.82\,\text{s}$
     (b) $x = 434.5\,\text{m}$

3.13 (a) $t = 0.25\,\text{s}$
     (b) $v_x = 12.13\,\text{m/s}$

3.14 $s = 85.6\,\text{m}$

3.15 (a) $h = 12.4\,\text{m}$
     (b) $h_{max} = 26.3\,\text{m}$
     (c) $t = 3.18\,\text{s}$

3.16 (a) $h = 10.7\,\text{m}$
     (b) $h_{max} = 21\,\text{m}$
     (c) $t = 2.96\,\text{s}$

3.17 $x = 66.2\,\text{m}$

3.18 $\theta = 80.9°$

Chapter 4

4.1 $p = 41.67\,\text{kN s}$

4.2 (a) $\mathbf{p} = 6.25\,\text{MN s}$ at $20°$
    (b) $\mathbf{p}_x = 5.87\,\text{MN s}$
    (c) $\mathbf{p}_y = 2.14\,\text{MN s}$

4.3 $\mathbf{H} = 4.5\,\text{kN m s}$, sense given by RH
    screw rule

4.4 $p = 1.04\,\text{kN s}, L = 83.3\,\text{kN m s}$

4.5 (a) $v = 8.25\,\text{m/s}$
    (b) $L = 160\,\text{N m s}$

4.6 $\Delta p = -20\,\text{N s}$ and $+20\,\text{N s}$

4.7 $e = 0.706$

4.8 $v = 1.5\,\text{m/s}$

4.9 $v_a = 0,\ v_b = 1\,\text{m/s}$

4.10 (a) $e = 2/3$
        (b) $p_a = 9\,\text{N s},\ p_b = 0$
        (c) $\Delta p = -7.5\,\text{N s}$
        (d) $m_b = 3\,\text{kg}$

4.11 (a) $v = 6.26\,\text{m/s}$
        (b) $v = 3.76\,\text{m/s}$
        (c) $h = 0.72\,\text{m/s}$

4.12

Bounce	h *(mm)*
1	720
2	260
3	93
4	330×10^{-3}
5	120×10^{-3}
6	43.5×10^{-3}
7	15.7×10^{-3}

4.13 $e = 0.733$
        $v_a - 1.03\,\text{m/s at } 47°$

4.16 (a) $\Delta p = -20.8\,\text{kN s}$
        (b) $F = 2.08\,\text{kN}$

4.17 $M = 3.13\,\text{kN m}$

4.18 (a) $v = 13.09\,\text{m/s}$
        (b) $\mathbf{a}_c = 34.27\,\text{m/s}^2$ radially
              inwards
        (c) $\mathbf{F} = 1.8\,\text{kN}$

4.19 (a) $\mathbf{a}_c - 1.97\,\text{km/s}^2$ radially
              inwards
        (b) $\mathbf{F} = 3.95\,\text{kN}$ radially outwards

4.20 (a) $\alpha = -20.4\,\text{rad/s}^2$
        (b) $T = -347.1\,\text{N m}$
        (c) 171 rev

4.21 $v = 19.5\,\text{km/h}$
        $\Delta p = 1075\,\text{N s}$

Chapter 5

5.1 $W = 200\,\text{J}$

5.2 $W = 1.5\,\text{kJ}$

5.3 $W = 333.8\,\text{J}$

5.4 $E_p = 245\,\text{MJ}$
        $E_k = 118\,\text{MJ}$

5.5 $k = 500\,\text{kN/m}$
        $E_e = 256\,\text{J}$

5.6   (a) $\omega = 53.33\,\text{rad/s}$
        (b) $E_k = 8.53\,\text{kJ}$
        (c) $E_k = 12.53\,\text{kJ}$

5.7   (b) (i) $4.91\,\text{kN/m}$
              (ii) $4.97\,\text{J}$

5.8 $v = 22.15\,\text{m/s}$

5.9 $v = 10.8\,\text{m/s},\ \omega = 43.2\,\text{rad/s}$

5.10 $h = 286.7\,\text{m}$

5.11 (a) $E_e = 0.625\,\text{J}$
        (b) $v = 2.24\,\text{m/s}$

5.12 (a) $\delta_0 = 58.9\,\text{mm}$
        (b) $E_p = 39.24\,\text{J}$
        (c) $v = 6.26\,\text{m/s}$
        (d) $\delta = 122.8\,\text{mm}$

5.13 (b) $33.33\,\text{m}$

5.14 (a) $v = 0.952\,\text{m/s}$
        (b) $\Delta E_k = 381\,\text{J}$

5.15 $v = 8.29\,\text{m/s}$

5.16 $E_k = 93.75\,\text{J}$

5.17 (a) $v = 39.62\,\text{m/s}$
        (b) $s = 56.6\,\text{m}$

5.19 $T = 23.9\,\text{N m}$

5.20 $P = 16.7\,\text{MW}$

5.21 (a) $T_{\text{ave}} = 53.13\,\text{N m}$
        (b) $P_{\text{ave}} = 27.82\,\text{W}$

Chapter 6

6.2   (a) $F_x = 3.63\,\text{kN}$
        (b) $F_y = 6.21\,\text{kN}$
        (c) $\mathbf{F} = 7.2\,\text{kN at } 59.7°$
        (d) $\mathbf{R} = 7.2\,\text{kN at } 239.7°$

6.3 $\mathbf{R} = 14\,\text{kN at } 78°$

6.4 $\mathbf{R}_{\text{wall}} = 71\,\text{N at } 180°$
        $\mathbf{R}_{\text{floor}} = 255\,\text{N at } 73.9°$
        $\mu = 0.29$

6.5 $R_{\text{wall}} = 70\,\text{N}$ at 163.3°
$R_{\text{floor}} = 234\,\text{N}$ at 73.3°

6.6 $T_{\text{rope}} = 2452.5\,\text{N}$
$R_{\text{fixing}} = 2452.5\,\text{N}$ at 220°
$R_{\text{ceiling}} = 4445\,\text{N}$ at 65°

6.7 (a) $T = 294.3\,\text{N}$
(b) $T = 294.3\,\text{N}$
$R = 224\,\text{N}$ (top left and bottom right rollers)
$R = 0$ (top right and bottom left rollers)

6.8 $R = 3.3\,\text{kN}$ and $3.95\,\text{kN}$

6.9 (a) $R = 2084.6\,\text{N}$
(c) inboard wheels, $R = 1916\,\text{N}$
outboard wheels, $R = 2253\,\text{N}$
(d) $\mu = 0.177$

6.10 (a) $\mathbf{R} = 16.5\,\text{kN}$ at 44.8°
(b) $\mu = 1$

6.11 $\theta = 21.8^{\circ}$

6.12 $R = 500\,\text{N}$, $Mo = 125\,\text{N}\,\text{m}$

6.14 $\omega = 0.29\,\text{rad/s}$
or $2.8\,\text{rev/min}$

6.15 (a) $v = 22.15\,\text{m/s}$
(b) $v = 21.4\,\text{m/s}$

6.17 (a) $T_b = 22.5\,\text{N}\,\text{m}$
(b) $\alpha = 0.6\,\text{rad/s}$

6.18 (a) $\mu = 0.06$
(b) $T_b = 23\,\text{N}\,\text{m}$
(c) $F = 409\,\text{N}$
(d) $W = 52.4\,\text{kJ}$

6.19 (a) $a = 0.67\,\text{m/s}^2$
(b) $T_{\text{cable}} = 21.5\,\text{kN}$
(c) $T = 5.38\,\text{kN}\,\text{m}$
(d) $P = 21.53\,\text{kW}$

Chapter 7

7.1 $m = 15.86\,\text{kg}$

7.2 $V = 0.05\,\text{m}^3$

7.3 (a) $T = 373\,\text{K}$
(b) $T = 273\,\text{K}$
(c) $T = 255.2\,\text{K}$

7.4 $\dot{V} = 4.9 \times 10^{-3}\,\text{m}^3/\text{s}$, $\dot{m} = 3.93\,\text{kg/s}$

7.5 (a) $p = 98.1\,\text{kN/m}^2$
(b) $p = 198.1\,\text{kN/m}^2$

7.6 $T = 0\,\text{K}$, absolute zero

7.7 $F = 25.1\,\text{N}$

7.8 $Q = 0.75\,\text{MJ}$, $P = 5\,\text{kW}$

7.9 (a) $p = 550\,\text{kN/m}^2$
(b) $p = 110.7\,\text{kN/m}^2$
(c) $p = 247.2\,\text{kN/m}^2$

7.10 (a) $m = 0.34\,\text{kg}$
(b) $p = 5.7\,\text{bar}$

7.11 (a) and (c)

7.12 (b) and (d)

7.13 (c)

7.14 Yes

7.15 $F = 6.26\,\text{mN}$

7.16 $V = 1.25\,\text{m}^3$

7.17 (a) $\dot{m} = 6.38\,\text{kg/s}$
(b) $\dot{m} = 4.38\,\text{kg/s}$
(c) 6.38×10^{-3}, 2×10^{-3} and $4.38 \times 10^{-3}\,\text{m}^3/\text{s}$

7.18 (a) $\dot{V} = 0.016\,\text{m}^3/\text{s}$
(b) $\dot{m} = 2\,\text{kg/s}$

7.19 (a) $d = 32\,\text{mm}$ and $29\,\text{mm}$
(b) $\dot{m} = 17.6\,\text{kg/s}$
(c) $d = 43.2\,\text{mm}$

7.20 $F = 76.7\,\text{N}$

7.22 $\mathbf{F} = 574\,\text{N}$ at -67.5°

7.23 2

7.25 $v = 12.45\,\text{m/s}$

7.26 $p = 1.83\,\text{bar}$

7.27 $\dot{m} = 0.66\,\text{kg/s}$

7.28 $\dot{m} = 0.66\,\text{kg/s}$

7.29 $\dot{m} = 2.96\,\text{kg/s}$

Chapter 8

8.1 (a) $R_L = 40\,\text{kN}$
$Mo = -100\,\text{kN m}$
(b) $R_L = 800\,\text{kN}$
$Mo = -2\,\text{kN m}$
(c) $R_L = 275\,\text{N}$
$R_R = 225\,\text{N}$
(d) $R_L = 22.2\,\text{kN}$
$R_R = 17.8\,\text{kN m}$
(e) $R_L = 1.33\,\text{kN}$
$R_R = 8.67\,\text{kN m}$

8.2 (a) $M = -100\,\text{kN m}$ at support
(b) $M = -2\,\text{kN m}$ at support
(c) $M = 275\,\text{N m}$, 1 m from LH end
(d) $M = 31.21\,\text{kN m}$, 2.21 m from LH end
(e) $M = -6\,\text{kN m}$, at RH support

8.3 $R_A = 5\,\text{kN}$ at $+90°$
$R_B = 5\,\text{kN}$ at $+90°$

8.4 $F_{BH} = 0$
$F_{AB} = 5.77\,\text{kN}$ (compressive)
$F_{AH} = 2.89\,\text{kN}$ (tensile)

8.5 $F_{AH} = F_{EF} = F_{HG} = F_{FG} = 2.89\,\text{kN}$ (tensile)
$F_{DE} = F_{AB} = 5.77\,\text{kN}$ (compressive)
$F_{BG} = F_{DG} = 5.77\,\text{kN}$ (tensile)
$F_{BC} = F_{DC} = 5.77\,\text{kN}$ (compressive)
$F_{CG} = F_{BH} = F_{DH} = 0$

8.7 $F_{AB} = 5\,\text{kN}$ (tensile
$F_{AG} = 0$
$F_{GB} = 0$

8.8 $F_{BC} = 7.08\,\text{kN}$ (tensile)
$F_{BF} = 5\,\text{kN}$ (compressive)
$F_{GF} = 0$

8.9 $F_{CD} = 7.08\,\text{kN}$ (tensile)
$F_{FD} = 5\,\text{kN}$ (tensile)
$F_{FE} = 7.08\,\text{kN}$ (compressive)

$F_{DE} = 0$
$R_D = 11.2\,\text{kN}$ at $18.4°$
$R_E = 7.08\,\text{kN}$ at $180°$

Chapter 9

9.1 (a) $A = 263.9 \times 10^{-6}\,\text{m}^2$
(b) $\sigma = 75.8\,\text{MN/m}^2$
(c) $\varepsilon = 362.6 \times 10^{-6}$

9.2 (a) $\tau = 1.6\,\text{MN/m}^2$
(b) $\gamma = 61.5 \times 10^{-6}$

9.3 (a) $\varepsilon = 180 \times 10^{-6}$
(b) $\Delta l = 0.36\,\text{mm}$

9.4 (a) $\sigma = 28.3\,\text{MN/m}^2$
(b) $\varepsilon = 404 \times 10^{-16}$
(c) $\Delta l = 570 \times 10^{-6}\,\text{m}$

9.5 (a) $\sigma = -1.27\,\text{MN/m}^2$
(b) $\Delta l = -12.15 \times 10^{-6}\,\text{m}$

9.6 (a) $T = 85.5\,°\text{C}$
(b) $Q = 411\,\text{kJ}$

9.7 (a) $\Delta T = 232\,°\text{C}$
(b) $\Delta T = -232\,°\text{C}$

9.8 $\varepsilon = 1 \times 10^{-3}$, $F = 1.48\,\text{kN}$

9.9 $E = 370\,\text{GN/m}^2$
$\sigma_{0.1\%} = 650\,\text{MN/m}^2$

9.10 (a) $F = 314.2\,\text{kN}$
(b) $\Delta l = 1.25 \times 10^{-3}\,\text{m}$

9.11 (a) $\sigma = 35.4\,\text{MN/m}^2$
(b) $SF = 9$

9.12 (b), (c) and (d)

9.13 (a)

9.14 $d = 6.74\,\text{mm}$

9.15 $F = 400\,\text{kN}$

9.16 (a) $\sigma = -75.2\,\text{MN/m}^2$
(b) $R = 47\,\text{kN}$

Index